METHODS IN MOLECULAR BIOLOGY

For further volumes:
http://www.springer.com/series/7651

Arthritis Research

Methods and Protocols

Second Edition

Edited by

Shunichi Shiozawa

Department of Rheumatology, Kyushu University, Beppu, Japan

Editor
Shunichi Shiozawa
Department of Rheumatology
Kyushu University
Beppu, Japan

ISSN 1064-3745 ISSN 1940-6029 (electronic)
ISBN 978-1-4939-0403-7 ISBN 978-1-4939-0404-4 (eBook)
DOI 10.1007/978-1-4939-0404-4
Springer New York Heidelberg Dordrecht London

Library of Congress Control Number: 2014934695

Printed on acid-free paper

Humana Press is a brand of Springer
Springer is part of Springer Science+Business Media (www.springer.com)

Preface

It is my pleasure to provide the readers a new volume of the Methods in Molecular Biology, which deals with current techniques for the research of arthritis and related conditions. Nowadays, information can be easily shared by anyone in the world so quickly without boundary, and so are the techniques for experimental medicine. Further, the results obtained from medical research have increasingly become accurate, quantitative, and reliable. Accordingly, the experimental techniques described here should be brand new, but, at the same time, one should also note that even such techniques are based on chemistry and biochemistry. In this regard, therefore, I must first thank all the authors, because they realized what the methods in molecular biology sought to be and did not hesitate to share their state-of-the-art techniques with us.

In this book, intravital multiphoton microscopy technique is introduced to visualize bone resorption of mature osteoclasts in living bone marrow and joints, which is a novel technique enabling the study of dynamical movements of cells in vivo not only for bone resorption but also for arthritis and systemic autoimmunity. The techniques for evaluating exhausted CD8 T cell and for studying nucleic acid sensors and their effects are also introduced. Also described are the techniques for in vivo tetracycline-controlled transgenic mice and T cell receptor transgenic mice. The former is described in an overview form because how to select respective techniques is often keen to the success of experiments technique. The techniques to detect V(D)J recombination products and microRNA are described. Current techniques for isolating, making cell lines, and assaying the activity of iNKT cell and MAIT cell are also described. Since we have succeeded in inducing autoimmunity akin to systemic lupus erythematosus (SLE) in mice not prone to autoimmune diseases with almost 100 % efficiency, we describe this technique and how to isolate the particular T cell responsible for inducing autoimmunity. As a model of systemic sclerosis (scleroderma), the technique to make bleomycin-induced dermal fibrosis is described. Quantitative proteomic techniques are also available for screening novel biomarkers in sera, and this is described in relation to the biomarkers found in rheumatoid arthritis (RA). The method for genome-wide study is also overviewed and detailed from practical viewpoint. Then, from a viewpoint of data synthesis of the genetic data obtained in high-throughput genotyping era, the method using Bayesian-based genetic association analysis is described in a form of hypothesis-driven candidate gene association studies, which enables solid evaluation of the results. Finally, seemingly somewhat abrupt but novel techniques including time-of-flight secondary ion mass spectrometry assaying cartilage surface components and a label-free imaging technique using coherent anti-Stokes Raman scattering microscopy taking adipose tissue as an example are described. The latter enables visualization of unlabeled samples by using the novel microscopy.

Beppu, Japan *Shunichi Shiozawa*

Contents

Contributors

MARI AINOLA • *Department of Medicine/Invärtes medicin, Helsinki University Hospital, Helsinki, Finland; Department of Medicine, Institute of Clinical Medicine, University of Helsinki, Helsinki, Finland*
GHADA ALSALEH • *University of Strasbourg, Strasbourg, France*
PÉTER ANTAL • *Department of Measurement and Information Systems, Budapest University of Technology and Economics (BME), Budapest, Hungary*
JÉRÔME AVOUAC • *Université Paris Descartes, Service de Rhumatologie A et INSERM U1016, Hôpital, Cochin, Paris, France*
GONCALO BARRETO • *Institute of Clinical Medicine, University of Helsinki, Helsinki, Finland*
HUARD BERTRAND • *Team 8, Institut Albert Bonniot, La Tronche, France*
BENCE BOLGÁR • *Department of Measurement and Information Systems, Budapest University of Technology and Economics (BME), Budapest, Hungary*
EDIT I. BUZÁS • *Department of Genetics, Cell and Immunobiology, Semmelweis University, Budapest, Hungary*
ASAKO CHIBA • *Department of Immunology, Juntendo University School of Medicine, Tokyo, Japan*
SHINGO EIKAWA • *Okayama University, Shikata, Okayama, Japan*
ANDRÁS GEZSI • *Department of Genetics, Cell and Immunobiology, Semmelweis University, Budapest, Hungary*
JACQUES-ERIC GOTTENBERG • *University of Strasbourg, Strasbourg, France*
SACHIKO HIROSE • *Department of Pathology, Juntendo University School of Medicine, Tokyo, Japan*
MIKA HUKKANEN • *Institute of Biomedicine, Anatomy, University of Helsinki, Helsinki, Finland*
GÁBOR HULLÁM • *Department of Measurement and Information Systems, Budapest University of Technology and Economics, Budapest, Hungary*
YUKIYASU IIDA • *Toin Human Science and Technology Center, Toin University of Yokohama, Yokohama, Japan*
MASARU ISHII • *Department of Immunology and Cell Biology, Graduate School of Medicine and Frontier Biosciences, Osaka University, Suita, Osaka, Japan*
ANTTI ISOMÄKI • *Institute of Biomedicine, Anatomy, University of Helsinki, Helsinki, Finland*
EMILIA KAIVOSOJA • *Institute of Clinical Medicine, University of Helsinki, Helsinki, Finland*
TAKESHI KAMEYAMA • *Hokkaido University, Sapporo, Japan*
JUNICHI KIKUTA • *Department of Immunology and Cell Biology, Graduate School of Medicine and Frontier Biosciences, Osaka University, Suita, Osaka, Japan*
YRJÖ T. KONTTINEN • *Department of Medicine, Institute of Clinical Medicine, University of Helsinki, Helsinki, Finland*
MIKKO LILJESTRÖM • *Institute of Biomedicine, Anatomy, University of Helsinki, Helsinki, Finland*
ANDRÁS MILLINGHOFFER • *Department of Measurement and Information Systems, Budapest University of Technology and Economics, Budapest, Hungary*

SACHIKO MIYAKE • *Department of Immunology, Juntendo University School of Medicine, Tokyo, Japan*
YUMI MIYAZAKI • *Department of Rheumatology, Kyushu University Beppu Hospital, Beppu, Japan*
SHUSAKU MIZUKAMI • *Okayama University, Shikata, Okayama, Japan; Department of Immunology, Okayama University Graduate School of Medicine, Dentistry and Pharmaceutical Sciences, Okayama, Japan*
TETSUJI NAKA • *Laboratory for Immune Signal, National Institute of Biomedical Innovation, Osaka, Japan*
HIOROYUKI NISHIMURA • *Toin Human Science and Technology Center, Toin University of Yokohama, Yokohama, Japan*
MASAOMI OBATA • *Toin Human Science and Technology Center, Toin University of Yokohama, Yokohama, Japan*
MAREKI OHTSUJI • *Toin Human Science and Technology Center, Toin University of Yokohama, Yokohama, Japan*
ZSUZSANNA PÁL • *Department of Genetics, Cell and Immunobiology, Semmelweis University, Budapest, Hungary*
NORIO SAKAI • *Department of Molecular and Pharmacological Neuroscience, Institute of Biomedical and Health Sciences, Hiroshima University, Hiroshima, Japan*
PÉTER SÁRKÖZY • *Department of Measurement and Information Systems, Budapest University of Technology and Economics, Budapest, Hungary*
SATOSHI SERADA • *Laboratory for Immune Signal, National Institute of Biomedical Innovation, Osaka, Japan*
SHUNICHI SHIOZAWA • *Department of Rheumatology, Kyushu University, Beppu, Japan*
TOSHIKAZU SHIRAI • *Department of Pathology, Juntendo University School of Medicine, Tokyo, Japan*
TARVO SILLAT • *Institute of Clinical Medicine, University of Helsinki, Helsinki, Finland*
ANTTI SOININEN • *ORTONResearch Institute, Diamond Group, ORTON, Orthopaedic Hospital of the ORTON Foundation, Diamond Group, Helsinki, Finland*
SANJEEV K. SRIVASTAVA • *Department of Genetics, Cell and Immunobiology, Semmelweis University, Budapest, Hungary*
AKINORI TAKAOKA • *Hokkaido University, Sapporo, Japan*
KEN TSUMIYAMA • *Department of Rheumatology, Kyushu University, Beppu, Japan*
HEIICHIRO UDONO • *Okayama University, Shikata, Okayama, Japan; Department of Immunology, Okayama University Graduate School of Medicine, Dentistry and Pharmaceutical Sciences, Okayama, Japan*
KENICHI UTO • *Department of Laboratory Medicine, Kobe University, Kobe, Japan*

Chapter 1

Intravital Multiphoton Microscopy for Dissecting Cellular Dynamics in Arthritic Inflammation and Bone Destruction

Junichi Kikuta and Masaru Ishii

Abstract

Osteoclasts are giant bone-resorbing polykaryons that differentiate from mononuclear macrophage/monocyte-lineage hematopoietic precursors. They play critical roles not only in normal bone homeostasis (remodeling) but also in the pathogenesis of bone-destructive disorders such as osteoporosis and rheumatoid arthritis. However, how the activity of mature osteoclasts is regulated in vivo remains unclear. To answer this question, we recently developed an advanced imaging system to visualize living bone tissues with intravital multiphoton microscopy. Using this system, we succeeded in visualization of mature osteoclasts in living bones.

We herein describe the detailed methodology for visualizing bone resorption of mature osteoclasts in living bone marrow and joints using intravital multiphoton microscopy. This approach would be beneficial for studying the cellular dynamics in arthritic inflammation and bone destruction in vivo and would thus be useful for evaluating novel anti-bone-resorptive drugs.

Key words Intravital imaging, Multiphoton microscopy, Osteoclast, Bone resorption, Th17 cell

1 Introduction

Rheumatoid arthritis (RA) is a chronic autoimmune disease characterized by synovial joint inflammation and progressive cartilage/bone destruction [1]. Various cell types, such as macrophages, T/B lymphocytes, and synovial fibroblasts, are reportedly involved in the pathogenesis of chronic inflammation in RA [2, 3]. However, arthritic bone destruction is considered to be mediated mainly by enhanced activation of osteoclasts at inflammatory sites. Osteoclasts are a specialized cell subset with a bone-resorbing capacity and play a critical role in normal bone homeostasis (bone remodeling), degradation of bone matrices, and facilitation of new bone formation by osteoblasts [4]. To prevent RA-associated bone destruction, it is important to understand the cellular dynamics of osteoclastic bone resorption in vivo.

Shunichi Shiozawa (ed.), *Arthritis Research: Methods and Protocols*, Methods in Molecular Biology, vol. 1142, DOI 10.1007/978-1-4939-0404-4_1, © Springer Science+Business Media New York 2014

Because bone is the hardest tissue in the body, it is difficult and almost impossible to visualize the inner bone tissue in living animals. In the fields of bone and mineral research, cell morphology and structure in bone tissues can be analyzed by conventional methods such as micro-CT and histological analysis. These methods allow for the evaluation of cell shape and molecular expression but cannot observe living osteoclast movement with respect to blood flow circulation. Thus, how the bone-resorptive functions of mature osteoclasts are controlled in vivo remains unclear.

To answer this question, we utilized an advanced imaging system to visualize living bone tissues with intravital multiphoton microscopy that we originally established [5, 6]. Multiphoton (usually two-photon) excitation-based laser microscopy has some advantages over confocal microscopy: increased penetration depth (up to 10–1,000 μm) and reduced tissue damage. It has enabled the visualization of dynamic cell behavior in deep intravital tissues and quantitative analysis of their mobility and interactions [7–10]. Access of deep bone tissues is difficult because the infrared laser is readily scattered by calcium phosphate crystals in the bone matrix. However, in the mouse parietal bone, the distance from the bone surface to the bone marrow cavity is only ~80–120 μm, which is thin enough to allow for passage of infrared lasers. Thus, we selected this region as the observation site.

To identify mature osteoclasts with fluorescence microscopy, they must be fluorescently labeled. Fully differentiated osteoclasts form an extracellular compartment (resorption lacunae) between the plasma membrane (ruffled border) and the bone surface. A large number of vacuolar type H^+-ATPases (V-ATPase) are specifically expressed along the ruffled border membrane to maintain highly acidic conditions in the resorption pit [11, 12]. V-ATPase comprises multiple subunits, and each subunit has several isoforms. Among them, the a3 isoform of the a-subunit is preferentially and abundantly expressed in mature osteoclasts [13, 14]. To fluorescently label mature osteoclasts, we generated mice in which a3 subunit-GFP fusion proteins are expressed under the original promoter of the a3 subunit (a3-GFP knock-in mice) [15].

Using intravital multiphoton microscopy of calvaria bone tissues of a3-GFP knock-in mice, we succeeded in visualizing the in vivo behavior of living mature osteoclasts on the bone surface [16]. Because GFP is expressed as a fusion protein with the a3 subunit, GFP fluorescence not only serves as a marker for mature osteoclasts but also provides information on the subcellular distribution of V-ATPase in osteoclasts. We have identified different functional subsets of living mature osteoclasts, from static–bone resorptive to moving–non-resorptive, for the first time. Treatment with recombinant RANKL or bisphosphonate changed the composition of these populations as well as the total number of mature osteoclasts.

We also found that RANKL not only promotes the differentiation of osteoclasts but also regulates the bone-resorptive function of fully differentiated mature osteoclasts [16].

Furthermore, we revealed the interaction of mature osteoclasts with $CD4^+$ T helper 17 (Th17) cells, one of the $CD4^+$ helper T subsets. Th17 cell numbers increase in the synovial fluid of patients with RA and enhance bone destruction associated with osteoclasts [17, 18]. Th17 cells reportedly express RANKL on their surface [19], which is suggested to be important for osteoclastic bone destruction in arthritic joints. However, the RANKL expressed on the surface of Th17 possesses little ability to induce differentiation, and the practical function of RANKL and Th17 on bone erosion remained elusive. To investigate how Th17 cells control osteoclastic bone resorption in vivo, polyclonally differentiated Th17 cells were labeled with fluorescent dye and then adoptively transferred into a3-GFP mice. We found that RANKL-bearing Th17 cells control bone resorption of mature osteoclasts, demonstrating novel actions of Th17, which may thus be a novel therapeutic target in RA [16].

In this chapter, we describe the detailed methodology for visualizing Th17-regulated bone destruction by osteoclasts in living bone tissues. In addition to this bone imaging technique, we also describe the practical imaging of bone destruction in arthritic joints using intravital multiphoton microscopy.

2 Materials

2.1 Multiphoton Microscopy Setup (Fig. 1)

1. Upright multiphoton microscope (A1-MP; Nikon) (*see* **Note 1**).
2. Water immersion objective, 25× (APO: numerical aperture [NA], 1.0; working distance [WD], 2.0 mm; Nikon) (*see* **Note 2**).
3. Femtosecond-pulsed infrared laser (Chameleon Vision II Ti: Sapphire laser; Coherent) (*see* **Note 3**).
4. Non-descanned detector (NDD) with 2–4 channels.
5. Customized microscope stage.
6. Environmental chamber in which anesthetized mice are warmed to 37 °C by an air heater.

2.2 Anesthesia

1. Male or female a3-GFP knock-in mice [15].
2. Isoflurane (Escain) (Mylan).
3. Vaporizer (inhalation device) (Baxter; 2.5 % vaporized in an 80:20 mixture of O_2 and air).
4. O_2 bomb.
5. Anesthesia box and mask.

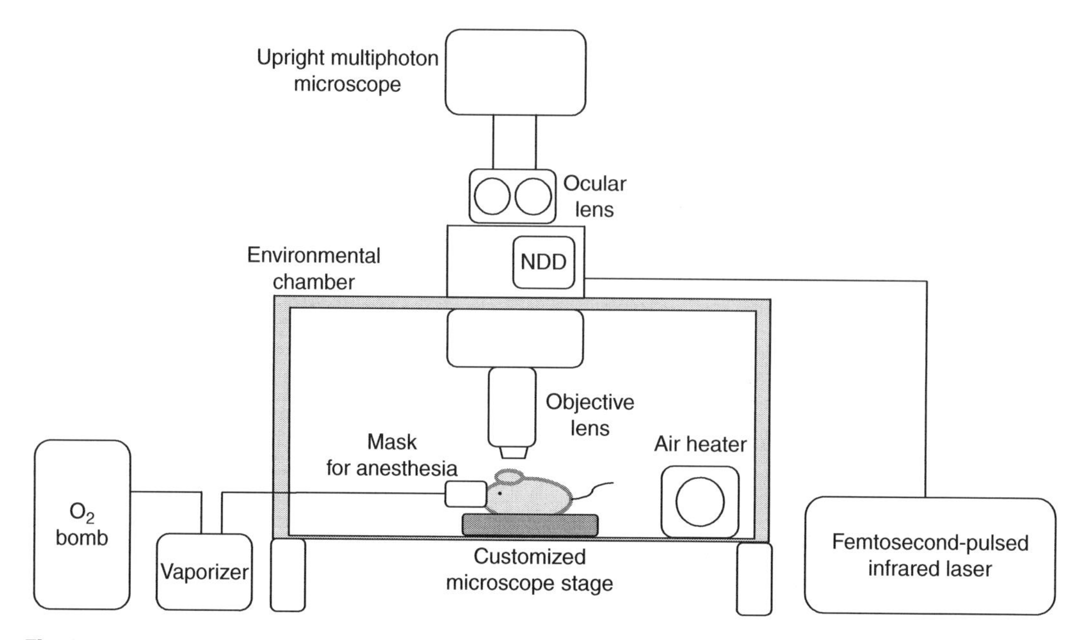

Fig. 1 View of multiphoton microscopy setup. During intravital multiphoton imaging, an anesthetized mouse is immobilized on the customized stage and warmed at 37 °C in the environmental chamber

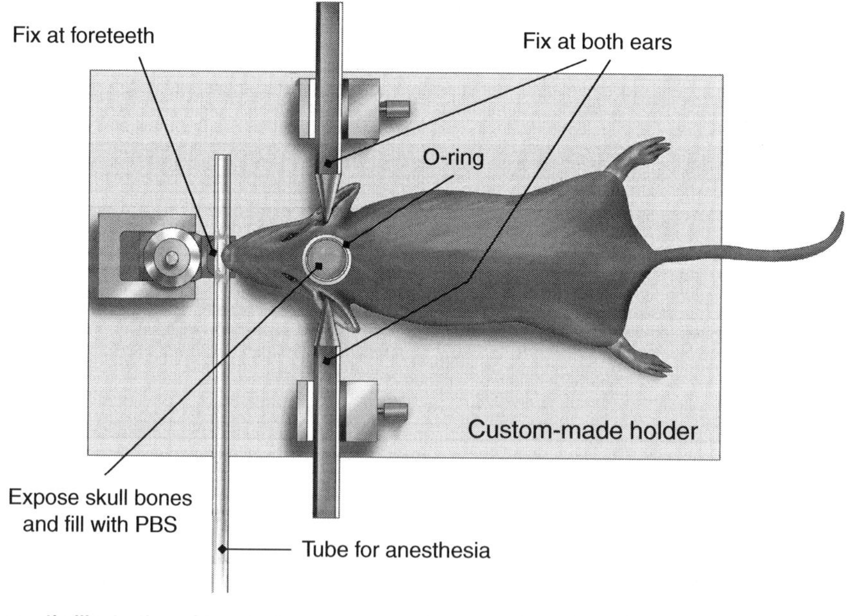

Fig. 2 Schematic illustration of how to fix a mouse on the stage for calvaria bone imaging. The mouse's head is immobilized with fixing at three points: fore-teeth and both ears. The O-ring is inserted into the incision of the skin and is filled with PBS

2.3 Preparation of Mice

1. Custom-made stereotactic holder (Figs. 2 and 4).
2. Shaver and hair-removal lotion (Epilat).
3. Iris scissors and tweezers for mouse operation.
4. O-ring: A 1.5-mL microtube is cut into a 2-mm-thick slice (*see* **Note 4**).
5. Instant adhesive and petrolatum or Difloil grease.
6. Phosphate-buffered saline (PBS) immersion buffer, pH 7.4.

2.4 Staining of Blood Vessels

1. Angiographic agent: 2 mg/mL of 70-kDa Texas Red-conjugated dextran (Molecular Probes) dissolved in PBS (*see* **Note 5**).
2. Several 29-G insulin syringes for intravenous injection.

2.5 Image Analysis

1. Image analysis software: Imaris (Bitplane) or Volocity (PerkinElmer).
2. After Effects (Adobe).

2.6 Preparation of Th17 Cells

1. Wild-type mice.
2. Dynabeads FlowComp Mouse $CD4^+$ Isolation Kit (Life Technologies).
3. Cytokines: Anti-CD3 mAb (Biolegend), anti-CD28 mAb (Biolegend), anti-IFN-γ mAb (R&D), anti-IL-4 mAb (R&D), and IL-23 (R&D).
4. RPMI medium with 10 % fetal bovine serum (FBS) containing 1 % penicillin and streptomycin.
5. Fluorescent dye: Celltracker Red CMTPX (Life Technologies) (*see* **Note 6**).
6. CO_2-independent medium (Life Technologies).
7. PBS, pH 7.4.
8. FBS.
9. Several 29-G insulin syringes.

2.7 Preparation of Collagen-Induced Arthritis

1. Heat-killed *Mycobacterium tuberculosis* strain H37Ra (BD Biosciences) for laboratory-prepared Freund's adjuvant (*see* **Note 7**).
2. Complete Freund's adjuvant (CFA; BD Biosciences) for commercially prepared Freund's adjuvant (*see* **Note 7**).
3. Incomplete Freund's adjuvant (IFA; BD Biosciences).
4. Chicken or bovine type II collagen (lyophilized form; Sigma).
5. Acetic acid.
6. Homogenizer.
7. One 26-G needle.

3 Methods

3.1 Intravital Multiphoton Imaging of Interactions of Osteoclasts with Th17 Cells

3.1.1 Preparation of Th17 Cells

1. Purify $CD4^+$ T cells from spleens and peripheral lymph nodes of wild-type mice using Dynabeads FlowComp Mouse $CD4^+$ Isolation Kit (Life Technologies) according to the manufacturer's instructions.
2. Culture $CD4^+$ T cells with a plate-bound 1 μg/mL anti-CD3 mAb and 1 μg/mL anti-CD28 mAb in RPMI medium containing 10 μg/mL anti-IFN-γ mAb, 10 μg/mL anti-IL-4 mAb, and 10 ng/mL IL-23 at 37 °C for 3 days. $CD4^+$ T cells can be differentiated into Th17 cells.
3. Collect and centrifuge the cells.
4. Aspirate the supernatant.
5. Resuspend cells gently in 2-mL CO_2-independent medium containing 15 μM CMTPX.
6. Incubate cells at 37 °C for 10 min.
7. Add 1 mL FBS, and incubate cells at 37 °C for another 3 min.
8. Add 7-mL CO_2-independent medium, and centrifuge the cells.
9. Wash cells with PBS twice.
10. Resuspend cells in 100 μL PBS.
11. Intravenously transfer the labeled cells to a3-GFP mice.
12. 2 h later, observe calvaria bone tissues using intravital multiphoton microscopy.

3.1.2 Intravital Multiphoton Imaging of Calvaria Bone Tissues

1. Start up the multiphoton microscope, and turn on the heater in the environmental chamber (*see* **Note 8**).
2. All procedures on mice are performed under anesthesia.
3. Shave the hair, and apply hair-removal lotion on top of the head of the mouse (*see* **Note 9**).
4. Cut the skin minimally with iris scissors for insertion of the O-ring.
5. Fix the O-ring on the parietal bone with adhesive and petrolatum or Difloil grease (*see* **Note 10**), which prevents leakage of PBS and fills the O-ring.
6. Immobilize the mouse on the custom-made stereotactic holder as tightly as possible to avoid drift secondary to respiration and pulsation (Fig. 2; *see* **Note 11**).
7. Focus on the bone marrow cavity at an appropriate depth, and look through ocular lenses with the help of a mercury lamp.
8. Change the light source from the mercury lamp to the Ti-sapphire laser and the optical path to the NDD. Set the zoom ratio, *z*-positions, interval time, and duration time using observation software attached to the microscope (Fig. 3).

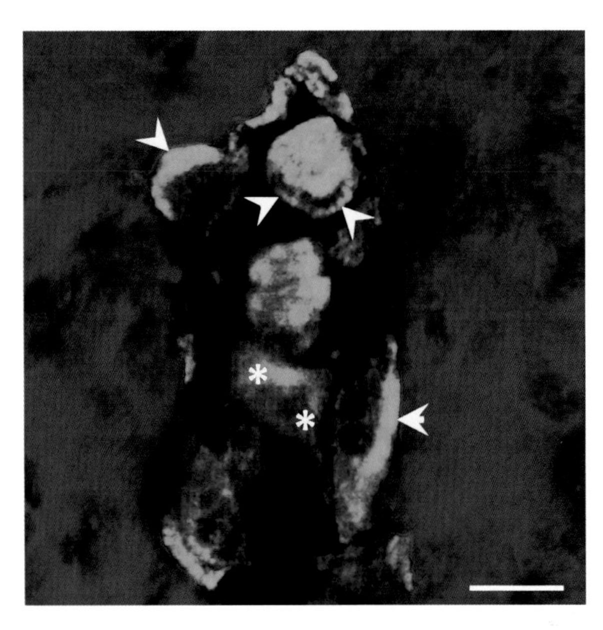

Fig. 3 Visualization of living mature osteoclasts on the bone surface by using intravital multiphoton microscopy. A representative image of calvaria bone imaging of a3-GFP mice in control conditions. *Green*, mature osteoclasts expressing GFP-fused V-type H^+-ATPase a3 subunit; *blue*, bone surface. Scale bar: 20 μm. *Arrowheads* and *asterisks* represent surface and cytoplasmic distribution of V-type H^+-ATPase a3 subunit, respectively

9. Analyze images by measuring cellular velocities, migration lengths, and contact times using image processing and analysis software.

3.2 Intravital Multiphoton Imaging of Arthritic Inflammation

3.2.1 Induction of Collagen-Induced Arthritis

1. Dissolve lyophilized type II collagen in 10 mM acetic acid to the desired concentration (2–4 mg/mL) by gentle stirring at 4 °C overnight (*see* **Note 12**).
2. Calculate the volumes of collagen and CFA required to make the emulsion, keeping their ratio at 1:1 (*see* **Note 13**).
3. Mix the emulsion with the homogenizer (*see* **Note 14**).
4. Slowly inject emulsified type II collagen in CFA (100-μL volume) intradermally into the tail.
5. 21 days later, inject emulsified type II collagen in IFA (100-μL volume) intradermally into the tail (*see* **Note 15**).
6. After arthritis development, observe arthritic joints using intravital multiphoton microscopy.

3.2.2 Intravital Multiphoton Imaging of Arthritic Joints

1. Start up the multiphoton microscope, and turn on the heater in the environmental chamber.
2. All procedures on mice are performed under anesthesia.

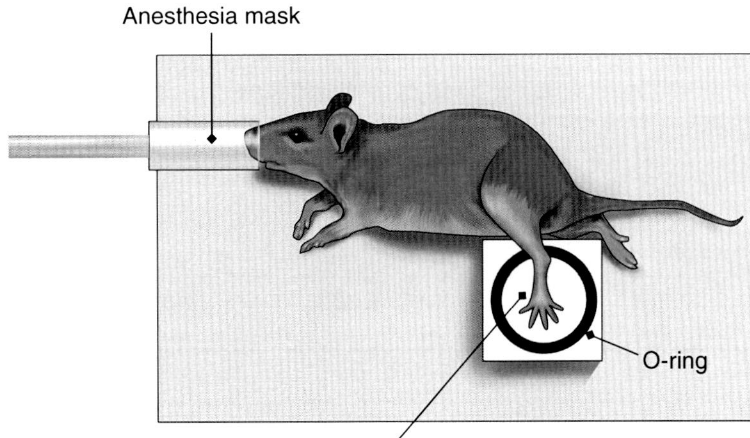

Fig. 4 Schematic illustration of how to fix a mouse on the stage for arthritic joint imaging. The mouse's hind limb is immobilized inside the O-ring which is filled with PBS

3. Shave the hair, and apply hair-removal lotion on the hind limb of the mouse.
4. Cut the skin and tendons minimally with iris scissors, and then expose the arthritic joints.
5. Immobilize the hind limb of the mouse on the custom-made holder (Fig. 4).
6. Fill the O-ring with PBS.
7. Intravenously inject 100 μL of 2-mg/mL 70-kDa Texas Red-conjugated dextran in PBS.
8. Focus on the joint cavity at an appropriate depth, and look through ocular lenses with the help of a mercury lamp.
9. Change the light source from the mercury lamp to the Ti-sapphire laser and the optical path to the NDD. Set the zoom ratio, *z*-positions, interval time, and duration time.
10. Analyze images using image processing and analysis software.

4 Notes

1. There are two types of microscopes: upright and inverted. Bone marrow can be observed using an inverted microscope. Multiphoton microscopy is also available from other microscope manufacturers (Leica Microsystems, Carl Zeiss, and Olympus).
2. Higher NA and longer WD are desirable for objective lenses.
3. A femtosecond-pulsed infrared laser is also available from Spectra-Physics (MaiTai).

4. The objective lens requires almost the same WD as that from the observation site. In addition, because the objective lens is water immersible, the substance must be filled with the same refractive index as that of water between the lens and observation site. The O-ring acts as both a spacer and a PBS reservoir.
5. We usually use a red fluorescent dye in this protocol because the target cells express GFP. If the target cells are red, fluorescein isothiocyanate (FITC)-conjugated dextran can be used. Because the blood vessels inside the bone marrow cavity have relatively high permeability, dextran with a molecular weight of more than 70 kDa should be used.
6. Many types of fluorescent dye are available.
7. A higher concentration of the heat-killed *M. tuberculosis* strain H37Ra is required to elicit a high incidence of arthritis.
8. It takes some time for the laser and the temperature to stabilize.
9. Remove as much hair as possible to avoid hair entering the visual field because it produces autofluorescence.
10. Avoid glue contamination in the visual field because some glues can produce autofluorescence.
11. Do not fasten too tightly because the mouse can be injured.
12. The procedure should be carried out at 4 °C using reagents and glassware that are prechilled to avoid denaturing type II collagen. Once the collagen has been dissolved to its desired concentration, it should be aliquoted and frozen at −70 °C.
13. While 100 μL of emulsion per mouse are required, extra emulsion must be produced because of losses during processing and the dead space in the syringes.
14. All reagents and the emulsion must be kept cold during the procedure.
15. A booster is sometimes used 21 days after the primary immunization to ensure induction of a high incidence of arthritis. The same concentration of type II collagen is emulsified in IFA for this immunization. The injection site is proximal to the primary one.

References

1. Firestein GS (2003) Evolving concepts of rheumatoid arthritis. Nature 423:356–361
2. Bromley M, Woolley DE (1984) Histopathology of the rheumatoid lesion. Identification of cell types at sites of cartilage erosion. Arthritis Rheum 27:857–863
3. Shimizu S, Shiozawa S, Shiozawa K, Imura S, Fujita T (1985) Quantitative histologic studies on the pathogenesis of periarticular osteoporosis in rheumatoid arthritis. Arthritis Rheum 28: 25–31
4. Teitelbaum SL (2000) Bone resorption by osteoclasts. Science 289:1504–1508
5. Ishii M et al (2009) Sphingosine-1-phosphate mobilizes osteoclast precursors and regulates bone homeostasis. Nature 458:524–528

6. Ishii M, Kikuta J, Shimazu Y, Meier-Schellersheim M, Germain RN (2010) Chemorepulsion by blood S1P regulates osteoclast precursor mobilization and bone remodeling in vivo. J Exp Med 207:2793–2798
7. Cahalan MD, Parker I, Wei SH, Miller MJ (2002) Two-photon tissue imaging: seeing the immune system in a fresh light. Nat Rev Immunol 2:872–880
8. Germain RN, Miller MJ, Dustin ML, Nussenzweig MC (2006) Dynamic imaging of the immune system: progress, pitfalls and promise. Nat Rev Immunol 6:497–507
9. Wang BG, Konig K, Halbhuber KJ (2010) Two-photon microscopy of deep intravital tissues and its merits in clinical research. J Microsc 238:1–20
10. Germain RN et al (2008) Making friends in out-of-the-way places: how cells of the immune system get together and how they conduct their business as revealed by intravital imaging. Immunol Rev 221:163–181
11. Blair HC, Teitelbaum SL, Ghiselli R, Gluck S (1989) Osteoclastic bone resorption by a polarized vacuolar proton pump. Science 245: 855–857
12. Nakamura H, Moriyama Y, Futai M, Ozawa H (1994) Immunohistochemical localization of vacuolar H(+)-ATPase in osteoclasts of rat tibiae. Arch Histol Cytol 57:535–539
13. Toyomura T, Oka T, Yamaguchi C, Wada Y, Futai M (2000) Three subunit a isoforms of mouse vacuolar H(+)-ATPase. Preferential expression of the a3 isoform during osteoclast differentiation. J Biol Chem 275: 8760–8765
14. Toyomura T et al (2003) From lysosomes to the plasma membrane: localization of vacuolar-type H+-ATPase with the a3 isoform during osteoclast differentiation. J Biol Chem 278: 22023–22030
15. Sun-Wada GH, Tabata H, Kawamura N, Aoyama M, Wada Y (2009) Direct recruitment of H+-ATPase from lysosomes for phagosomal acidification. J Cell Sci 122:2504–2513
16. Kikuta J et al (2013) Dynamic visualization of RANKL and Th17-mediated osteoclast function. J Clin Invest 123:866–873
17. Kong YY et al (1999) Activated T cells regulate bone loss and joint destruction in adjuvant arthritis through osteoprotegerin ligand. Nature 402:304–309
18. Takayanagi H et al (2000) T-cell-mediated regulation of osteoclastogenesis by signalling cross-talk between RANKL and IFN-gamma. Nature 408:600–605
19. Sato K et al (2006) Th17 functions as an osteoclastogenic helper T cell subset that links T cell activation and bone destruction. J Exp Med 203:2673–2682

Chapter 2

Monitoring Multifunctionality of Immune-Exhausted CD8 T Cells in Cancer Patients

Shingo Eikawa, Shusaku Mizukami, and Heiichiro Udono

Abstract

CD8 T cells play a critical role in the host defense against cancers and infectious diseases. However, the presence of antigen-specific CD8 T cells does not always imply that cancers and/or pathogens are efficiently eliminated in the body. Concerning this point, the recent studies suggest the concept of immune exhaustion of CD8 T cells, characterized by their decreased production of IL-2, TNFα, and IFNγ even after antigen stimulation. Thus, continuous stimulation of CD8 T cells by the persistent antigens results in immune exhaustion, which eventually causes immune tolerance against cancers and chronic infections. The identification of immune effector and/or exhausted CD8 T cells by monitoring multiple parameters including T cell exhaustion markers such as PD-1 and Tim-3 and intracellular cytokines is, therefore, crucial to understand the real-time, ongoing immune status. For this purpose, polychromatic flow cytometry is the most common and reliable tool to monitor T cell functions.

We describe here the method for detection of immune-exhaustion status of CD8 T cells from human peripheral blood mononuclear cells (PBMCs). By stimulation of PBMCs with PMA/ionomycin for 6 h, more than 1–2 % of total CD8 T cells are identified as positive in terms of multifunctionality, thus producing multiple cytokines—IL-2, TNFα, and IFNγ—at single-cell level in case of all healthy donors. By contrast, CD8 T cells from certain populations of cancer patients are significantly less effective; less than 0.5 % of CD8 T cells are positive in producing such multiple cytokines. The cutoff value around 0.5 % of CD8 T cells might distinguish patients who would receive beneficial effect by cancer vaccine from those who would not respond to the vaccine. Thus, the remaining capacity to produce multiple cytokines of CD8 T cells might be a critical parameter determining the outcome of cancer patients who receive various kinds of cancer vaccines. The method to monitor the state of multifunctionality of CD8 T cells, as described here, would become more important to understand the immune statues in cancers and chronic infectious diseases such as AIDS and malaria infections.

Key words CD8 T cells, Immune-exhaustion, Multifunctionality, PD-1, Tim-3

1 Introduction

The clinical trials of immunotherapy including cancer vaccine such as protein or long peptide of tumor antigen have been performed [1–3]. The vaccination induced robust humoral and cellular

Shunichi Shiozawa (ed.), *Arthritis Research: Methods and Protocols*, Methods in Molecular Biology, vol. 1142, DOI 10.1007/978-1-4939-0404-4_2,

immune response against vaccinated tumor antigen and some of endogenous antigens [4, 5]. However, in some cases peripheral blood T cell responses in vitro will not always be consistent with the predicted clinical efficacy of a vaccine. For improvement of the case, it is essential to obtain accurate and more information from patients.

T cell exhaustion is a dysfunctional state occurring in cancer and chronic inflammation [6]. It is defined by low proliferative capacity, low cytokine production, and upregulation of inhibitory receptors on T cells [7]. During exhaustion in chronic inflammation, loss of function occurs for each stage. In general, firstly, IL-2 production, high proliferative capacity, and cytotoxicity are lost. Secondly, the ability of TNFα production is lost at intermediate dysfunctional stages. In the final stage, T cells could not produce the large amounts of IFNγ and severe exhaustion leads to cell apoptosis [7].

Thus, in tumor immunity, an investigation to evaluate the effect of a therapy requires the selection of optimized methods for functional and/or dysfunctional T cell immunomonitoring during immune exhaustion. We have optimized here the procedures to detect CD8 T cell response in human and mouse system. In human, the expression of exhaustion markers, PD-1 and Tim-3, on cell surface and intracellular cytokines, IL-2, TNFα, and IFNγ were detected in peripheral blood mononuclear cells (PBMCs) after stimulation with appropriate concentration of PMA/ionomycin. In mouse, OT-I CD8 T cells were stimulated with dendritic cells pulsed with specific antigen peptide ($OVA_{257-264}$), and the produced multiple cytokines were detected using intracellular cytokine staining (ICS) for IL-2, TNFα, and IFNγ.

By using these procedures, one might determine a cutoff value by which they can predict the prognosis of cancer patients, concerning whether each patient receives beneficial effect or not prior to vaccination of interest. In our laboratory, the cutoff value was nailed down to around 0.5 % of total CD8 T cells. Thus, patients whose ability to produce triple cytokines over 0.5 % responded to the peptide vaccine, retrospectively, while less than 0.5 % never received the beneficial effects (unpublished observations).

The cutoff value, of course, should be determined by comparing that obtained from enough numbers of healthy donors in each institute or hospitals, thus, depending on each facility.

2 Materials

2.1 Culture Medium

1. Human cell culture medium: AIM V® Medium (Gibco, Tokyo, Japan).
2. Mouse cell culture medium: Weigh 0.11 g of sodium pyruvate, 3.9 mg 2-ME (Sigma-Aldrich), 60 mg Penicillin–Streptomycin (Sigma-Aldrich), 0.29 g L-Glutamine (Sigma-Aldrich) and dissolve in 1 l of RPMI medium 1640 (Wako).

3. Weigh 0.58 g of EDTA (Nacalai Tesque, Kyoto, Japan) and dissolve in 1 l of PBS (Gibco) containing 2 % FCS (Thermo, Waltham, MA, USA).

2.2 Cell Activation

PMA (Sigma-Aldrich)/ionomycin (Sigma-Aldrich).

2.3 mAbs for Cell Surface Molecule Staining for Human

1. Anti-human CD8 allophycocyanin (APC)-Cy7 (cyanine dye)-conjugated monoclonal antibody (anti-hCD8-APC-Cy7) (BioLegend, San Diego, CA, USA).
2. Anti-human PD-1 (Programmed Death-1; CD279)-phycoerythrin (PE)-Cy7-conjugated monoclonal antibody (anti-hPD-1-PE-Cy7) (eBiosciences, San Diego, CA, USA).
3. Anti-human Tim-3 (T cell immunoglobulin mucin-3) PerCP-conjugated monoclonal antibody (anti-hTim-3-PerCP) (R&D Systems, Minneapolis, MN, USA).

2.4 mAbs for Intracellular Cytokine Staining for Human

1. Anti-human IL-2 APC-conjugated monoclonal antibody (anti-hIL-2-APC) (BD Biosciences, San Jose, CA, USA).
2. Anti-human TNFα PE-conjugated monoclonal antibody (anti-hTNFα-PE) (BD Biosciences).
3. Anti-human IFNγ FITC-conjugated monoclonal antibody (anti-hIFNγ -FITC) (BD Biosciences).

2.5 mAbs for Cell Surface Molecule Staining for Mouse

Anti-mouse CD8 APC-Cy7-conjugated monoclonal antibody (anti-m CD8-APC-Cy7) (BD Pharmingen).

2.6 mAbs for Intracellular Cytokine Staining for Mouse

1. Anti-mouse IL-2 PE-Cy7-conjugated monoclonal antibody (anti-m IL-2-PE-Cy7) (BD Biosciences).
2. Anti-mouse TNFα PerCP-Cy5.5-conjugated monoclonal antibody (anti-m TNFα-PerCP-Cy5.5) (BD Pharmingen).
3. Anti-mouse IFNγ FITC-conjugated monoclonal antibody (anti-mIFNγ-FITC) (eBioscience).

2.7 Staining Buffer

Weigh 0.58 g of EDTA (Nacalai Tesque) and dissolve in 1 l of PBS (Gibco) containing 2 % FCS (Thermo).

2.8 Protein Transport Inhibitor

BD Golgi Stop (containing monensin) (BD Biosciences) (*see* **Note 1**).

2.9 Cell Fixation/Permeabilization Buffer

BD Cytofix/Cytoperm (BD Biosciences).

2.10 Perm/Wash Buffer

BD Perm/Wash (dilute 1:10 in distilled H_2O prior to use) (BD Biosciences).

2.11 Flow Cytometer

Use the models like BD FACSCantoII™ Flow Cytometer (BECTON DICKINSON), equipped with two lasers, having the ability to detect six or more colors.

3 Methods

3.1 In Vitro Stimulation of Cell Culture

Human PBMCs (0.5–2 × 10^6/well) are cultured in the presence of 50 ng/ml PMA/2 μM ionomycin and then treated with BD Golgi Stop for 4–6 h (*see* **Notes 2** and **3**) in a 5 % CO_2 atm at 37 °C. OT-I T cells (5 × 10^5/well) are cultured with 1 μM H-2K^b presentable OVA peptide (SIINFEKL)-pulsed dendritic cells for 8–10 h.

3.2 Harvest Cells and Cell Staining

In human system, harvest cells from in vitro-stimulated culture and resuspend in staining buffer. Wash cells twice in staining buffer, and the media is completely removed. Stain cells in 50 μl of staining buffer with the appropriate amount of a fluorochrome-conjugated mAb binding for CD8, PD-1, and Tim-3 for 30 min on ice. Wash cells twice in 1 ml staining buffer and resuspend (*see* **Note 4**) in 500 μl BD Cytofix/Cytoperm solution for 30 min on ice. Wash cells twice in 1 ml BD Perm/Wash. Stain cells in 50 μl of BD Perm/Wash solution (*see* **Note 5**) with the appropriate amount of an intracellular specific mAb for IL-2, TNFα, and IFNγ for 30 min on ice (*see* **Note 6**). Wash cells twice in BD Cytofix/Cytoperm solution and resuspend in 200 μl staining buffer.

In mouse system, harvest cells as in human cells. Wash cells twice in staining buffer, and the media is completely removed. Incubate the cells with CD8 staining mAb for 30 min on ice, wash twice in 1 ml staining buffer, and then resuspend with 500 μl BD Cytofix/Cytoperm solution for 30 min on ice. Wash the cells twice by 1 ml BD Perm/Wash. Incubate the cells in 50 μl of BD Perm/Wash solution with the appropriate amount of an intracellular specific mAb for IL-2, TNFα, and IFNγ for 30 min on ice, wash twice in BD Cytofix/Cytoperm solution, and then resuspend in 200 μl staining buffer.

3.3 Data Analysis

First, determine PMT voltages and adjust compensation for the instrument and fluorochromatic reagent panel. Create template for acquisition in the pattern of dot plots. Here three strategies of gating analysis are displayed by flowjo (TreeStar, Ashland, OR, USA). Figure 1 shows the production of each cytokine in exhausted (PD-1$^+$Tim-3$^-$, PD-1$^+$Tim-3$^+$, or PD-1$^-$ Tim-3$^+$) or non-exhausted (PD-1$^-$Tim-3$^-$) CD8 T cells. On the other hand, Fig. 2 shows the multifunctionality and dysfunctional state in cytokine-positive CD8 T cells (IL-2$^+$, TNFα$^+$, or IFNγ$^+$). Figure 3 shows the multifunctionality of OT-ICD8 T cells stimulated with or without $OVA_{257-264}$ peptide for 8 h.

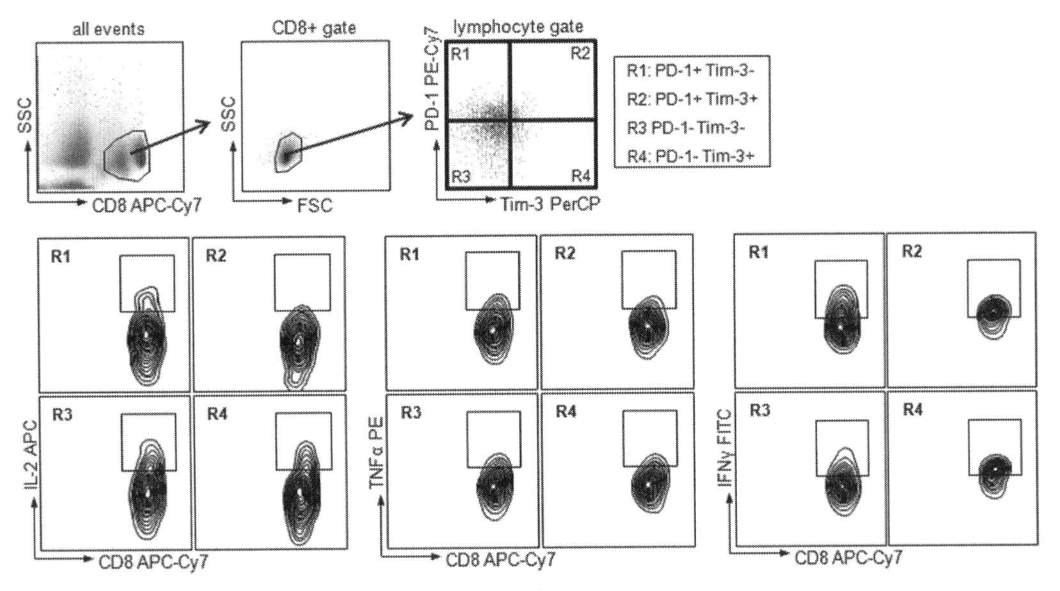

Fig. 1 A gating strategy for six-color staining for cell surface molecules (CD8, PD-1, and Tim-3) and intracellular cytokines (IL-2, TNFα, and IFNγ) in human PBMCs stimulated by PMA/ionomycin. The alternative gating by SSC and CD8, instead of FSC and SSC as usual, was used to ablate CD8-negative cells and debris. The expression of PD-1 and Tim-3 was determined after gating CD8^{+} lymphocytes. The cytokine-positive CD8 T cells was also defined on exhausted (PD-1^{+} Tim-3^{-}, PD-1^{+} Tim-3^{+}, or PD-1^{-} Tim-3^{+}) or non-exhausted (PD-1^{-} Tim-3^{-}) CD8 T cells. Of note that PD-1^{+}Tim-3^{+}CD8^{+} T cells are less effective in producing IL-2

4 Notes

1. Use 4 μl of BD Golgi Stop for every 6 ml of cell culture and mix thoroughly.
2. It is recommended that BD Golgi Stop should not be kept in cell culture for longer than 10–12 h.
3. Because stronger signal leads to elimination of surface markers such as PD-1 and Tim-3 and memory marker CD62L, it is important to determine optimized conditions in your experiment. Thus, the concentration of PMA/ionomycin should be twofold serial diluted (200, 100, and 50 ng/ml PMA/8, 4, and 2 μM ionomycin) and used for the assay in order to determine the best condition in your experiments.
4. Aggregation can be avoided by vortexing prior to treatment of the BD Cytofix/Cytoperm solution.
5. Because saponin-mediated cell permeabilization is a reversible process, it is important to keep the cells in the presence of saponin throughout the entire procedure of intracellular cytokine staining.
6. For reduction of the nonspecific response, it is recommended that cells were blocked by 2 % mouse or rat normal serum.

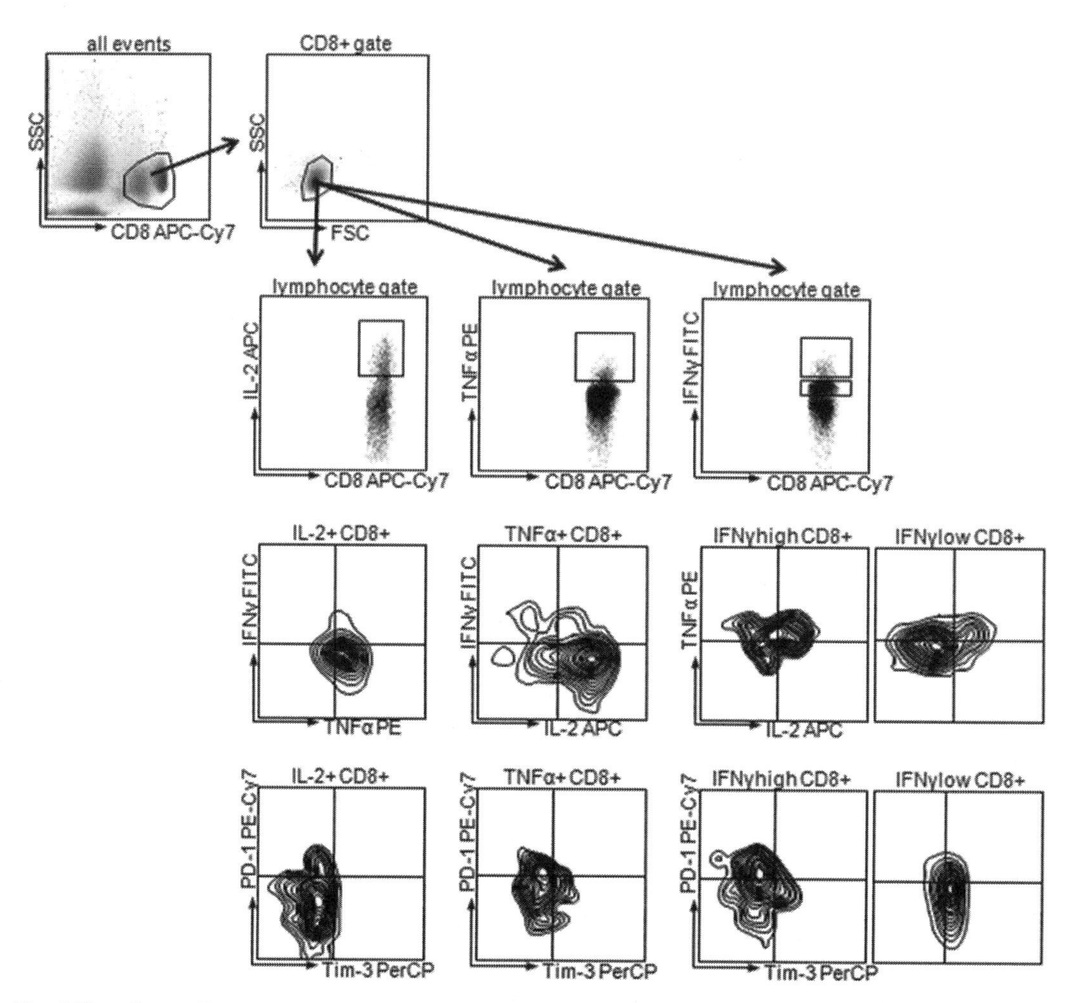

Fig. 2 The other gating strategy for six-color staining for molecules (CD8, PD-1, and Tim-3) and intracellular cytokines (IL-2, TNFα, and IFNγ) staining in human PBMCs stimulated with PMA/ionomycin. The alternative gating by SSC and CD8 was used to ablate CD8-negative cells and debris, as in Fig. 1. The cytokine-positive CD8 T cells were also defined by gating around CD8-positive lymphocytes. The multicytokine-positive CD8 T cells or the expression of PD-1 and Tim-3 were defined on each cytokine-positive cell. Of note that Tim-3^{+} cells are not effective in producing IL-2 and other cytokines

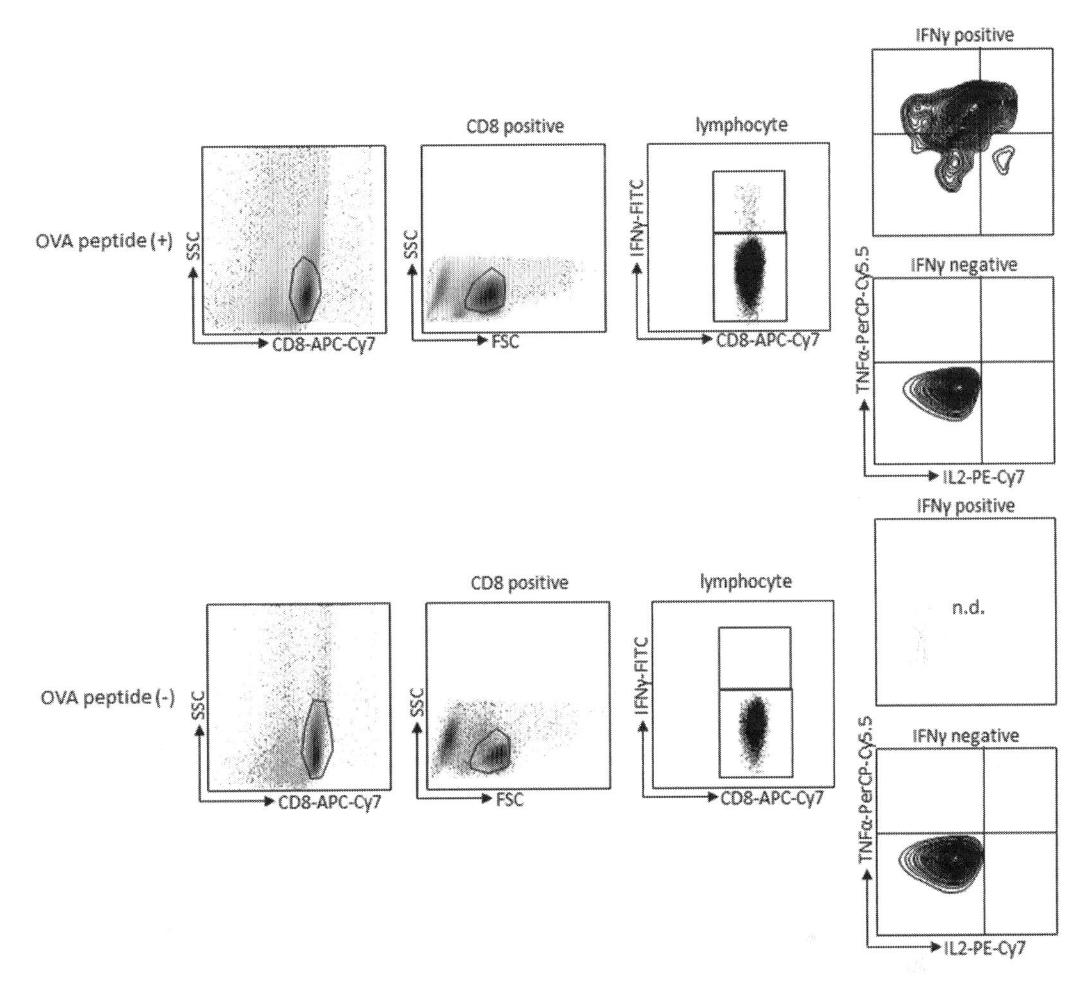

Fig. 3 A gating strategy for four-color staining including CD8 and intracellular cytokines (IL-2, TNFα, and IFNγ) on OT-I CD8 T cells that have been stimulated with OVA peptide (SIINFEKL)-pulsed dendritic cells. The alternative gating by SSC and CD8 was used to ablate CD8-negative cells and debris, as in Fig. 1. Thus, each cytokine-positive cell was identified after gating on CD8-positive T cells

References

1. Kakimi K, Isobe M, Uenaka A et al (2011) A phase I study of vaccination with NY-ESO-1f peptide mixed with Picibanil OK-432 and Montanide ISA-51 in patients with cancers expressing the NY-ESO-1 antigen. Int J Cancer 129:2836
2. Sabbatini P, Tsuji T, Ferran L et al (2012) Phase I trial of overlapping long peptides from a tumor self-antigen and poly-ICLC shows rapid induction of integrated immune response in ovarian cancer patients. Clin Cancer Res 18:6497
3. Odunsi K, Matsuzaki J, Karbach J et al (2012) Efficacy of vaccination with recombinant vaccinia and fowl pox vectors expressing NY-ESO-1 antigen in ovarian cancer and melanoma patients. Proc Natl Acad Sci U S A 109(15):5797–5802
4. Kawada J, Wada H, Isobe M et al (2012) Heteroclitic serological response in esophageal and prostate cancer patients after NY-ESO-1 protein vaccination. Int J Cancer 130:584
5. Jäger E, Karbach J, Gnjatic S et al (2006) Recombinant vaccinia/fowl pox NY-ESO-1 vaccines induce both humoral and cellular NY-ESO-1-specific immune responses in cancer patients. Proc Natl Acad Sci U S A 103(39): 14453–14458
6. Pardoll DM (2012) The blockade of immune checkpoints in cancer immunotherapy. Nat Rev Cancer 12:252
7. Wherry EJ (2011) T cell exhaustion. Nat Immunol 12:492

Chapter 3

Characterization of Innate Immune Signalings Stimulated by Ligands for Pattern Recognition Receptors

Takeshi Kameyama and Akinori Takaoka

Abstract

The innate immunity is an essential step as the front line of host defense, and its aberrant activation particularly in response to nucleic acids is closely related to the pathogenesis of autoimmune and inflammatory diseases. Characterization of the innate immune signalings may provide a pathophysiological insight for better understanding of human diseases. Nucleic acid-mediated activation of pattern recognition receptors triggers the activation of two major intracellular signaling pathways, which are dependent on NF-κB and interferon regulatory factors, transcriptional factors. This leads to the subsequent induction of inflammatory cytokines and type I and III interferons. In this chapter, we first overview the representative families of nucleic acid sensors and their ligands and then show the fundamental techniques for extracellular or intracellular stimulation with these nucleic acid ligands and for detection of innate immune response, that is, IFN and proinflammatory cytokine induction, as assessed by luciferase assay, quantitative RT-PCR (qRT-PCR), and enzyme-linked immunosorbent assay.

Key words Innate immunity, Pattern recognition receptors, Interferons, Inflammatory cytokines, Nucleic acids

1 Introduction

Microbial infection is an etiological and pathological factor to modify human diseases in a variety of aspects. Innate immune system acts as a front line of host defense against infection by microbes, such as viruses, bacteria, and fungi. To protect from invasion of these pathogens, host cells need to discriminate between self and nonself molecules. In this step, pattern recognition receptors (PRRs) have an essential role in the recognition of nonself molecules. PRR-mediated signalings result in the activation of NF-κB and interferon regulatory factors (IRFs) and the subsequent induction of various inflammatory cytokines, such as IL-6 and TNF, and type I and type III interferons (IFNs), respectively. Other PRRs trigger proinflammatory responses via the TLR- and IRF-independent pathway mediated by the inflammasome, which leads

Shunichi Shiozawa (ed.), *Arthritis Research: Methods and Protocols*, Methods in Molecular Biology, vol. 1142, DOI 10.1007/978-1-4939-0404-4_3, © Springer Science+Business Media New York 2014

to the maturation of the interleukin-1 family of cytokines, such as IL-1β and IL-18. These innate immune signalings contribute to not only inducing the early host response against pathogens but also linking to adaptive immune system in the late phase of infection. The coordinated responses by innate and adaptive immune systems are essential for efficient elimination of invading pathogens [1–4].

Germline-encoded PRRs comprise at least five families (Table 1): Toll-like receptors (TLRs), RIG-I-like receptors (RLRs), NOD-like receptors (NLRs), AIM2-like receptors (ALRs), and C-type lectin receptors (CLRs) [3]. These PRRs usually recognize molecular pattern associated with microbes (i.e., pathogen-associated molecular patterns; PAMPs) as shown in Table 1. On the other hand, there have been accumulating reports, which show that PRRs play an important role in sensing not only microbe-derived, nonself molecules but also self molecules in some cases such as trauma, ischemia, cancer, and other settings of tissue damage. Increased serum levels of these endogenous molecules (e.g., high-mobility group box 1; HMGB1, amyloid A protein, hyaluronic acid, uric acid, and self-DNA), the so-called danger-associated molecular patterns (DAMPs), which can activate the innate immune response, are associated with inflammatory and autoimmune diseases, including inflammatory bowel diseases, arthritis, and systemic lupus erythematosus [5, 6]. In addition, even among these PAMPs (Table 1), in particular, several types of "nucleic acids" can be detected in mammalian host cells. In fact, there are several reports showing that mammalian DNAs can evoke type I IFN response, which is one of the hallmarks of innate immune response [7, 8]. It has been thought that host cell-derived nucleic acids are spatially regulated or degraded so that they become incompetent to be sensed by PRRs. In this regard, one can easily envisage that dysregulation of nucleic acid sensor-mediated signalings may possibly lead to autoimmune abnormalities and inflammatory diseases. Characterization of these signalings may provide important insights into disease pathogenesis. In this chapter, we focus on nucleic acid innate sensors and their signalings to induce cytokine response. We introduce a brief summary of nucleic acid sensors in the innate immunity system and describe detailed protocols for preparation of nucleic acid ligands for those DNA/RNA sensors, cell stimulation, and detection of nucleic acid-induced cytokine responses.

Nucleic acids, one of the major PAMPs, are recognized by their specific sensors, which can be categorized into two types in terms of their subcellular localization: (a) membrane-associated type and (b) cytosolic type [9]. Several members of the TLR family are well studied as a representative of membrane-associated-type nucleic acid sensors. As for RNA recognition, TLR3 and TLR7/8 detect double-stranded RNA (ds-RNA) and single-stranded RNA (ss-RNA), respectively [3]. Recently it has been shown that in mice bacterial 23S ribosomal RNA is sensed by TLR13, which is not

Table 1
Pattern recognition receptors and their ligands

Family	PRRs	Ligands
TLRs	TLR1/TLR2	Triacyl lipopeptides (Pam3CSK4)
		Peptidoglycans
		Lipopolysaccharides
	TLR2/TLR6	Diacyl lipopeptides (FSL-1)
		Lipoteichoic acid
	TLR3	ds-RNA, poly(rI:rC)
		tRNA, siRNA
	TLR4	Lipopolysaccharides (LPS)
		Paclitaxel
	TLR5	Flagellin
	TLR7/8	ss-RNA
		Imidazoquinolines (R848)
		Guanosine analogs (loxoribine)
	TLR9	CpG-ODN
		(ODN2006; human, ODN1826; mouse)
	TLR11	Profilin-like molecule
	TLR13	23S ribosomal RNA (ORN Sa19)
RLRs	RIG-I	ds-RNA (short)
		3pRNA
		B-DNA
	MDA5	ds-RNA (long)
		Poly(rI:rC)
ALRs	AIM2	B-DNA
	IFI16	dsVACV 70 mer
NLRs	NOD1	Diaminopimelic acid, iE-DAP
		Tri-DAP, murabutide
	NOD2	MDP
		ss-RNA
	NLRP3	ATP, uric acid crystals, RNA, DNA, MDP
CLRs	Dectin-1	β-Glucan
	Dectin-2	β-Glucan
	MINCLE	SAP130
Others	DAI	ds-DNA
	DDX41	ds-DNA
	STING	Cyclic di-GMP, cyclic di-AMP
		Cyclic GMP-AMP (cGAMP)

expressed in humans [10]. On the other hand, TLR9 recognizes specific DNA sequences that contain unmethylated CpG motifs, which are commonly found in the genomes of bacteria and viruses [11]. The engagement of these TLRs with nucleic acid ligands results in the activation of intracellular signaling pathways mediated by NF-κB and IRFs, leading to the gene induction of proinflammatory cytokines, type I/III IFNs, chemokines, and costimulatory molecules, involved in antimicrobial immunity.

As cytosolic type RNA sensors, at least three DExD–H-box RNA helicases, retinoic acid inducible gene I (RIG-I), melanoma differentiation-associated gene 5 (MDA5), and laboratory of genetics and physiology 2 (LGP2), are included in the RIG-I-like receptor family [12, 13]. RIG-I and MDA5 are well-characterized members and key PRRs for the detection of positive- and negative-stranded RNA viruses that infect in the cytoplasmic space. Particularly, RIG-I has an important role in triggering responses to many viruses, such as those of the orthomyxovirus family (influenza A virus), paramyxovirus family (measles, mumps, and Sendai virus), hepatitis C virus, and Japanese encephalitis virus, most of which are causative agents for infectious diseases in human [14]. RIG-I targets RNA containing 5′-triphosphate modification (3pRNA) and short ds-RNA (<~1 kb) [15, 16], whereas MDA5 recognizes longer ds-RNA (>~2 kb) [13, 17, 18]. Interestingly, RIG-I also functions to indirectly detect cytosolic AT-rich double-stranded DNAs (ds-DNAs) (also known as B-form DNA; B-DNA) by its recognition of 3pRNA that is generated through the transcription by DNA-dependent RNA polymerase III [19, 20]. It is further reported that NOD2 (nucleotide-binding oligomerization domain-containing protein 2), which is known to be one of the NLR family members and a sensor for bacterial peptidoglycans, also senses virus-derived ss-RNAs in the cytoplasm, in an RLR-independent manner during the early stage of infection [21]. As shown in Fig. 1, RNA ligand binding results in the association of RLRs and NOD2 with the adaptor protein MAVS (mitochondrial antiviral signaling protein; also known as IPS-1, VISA, or Cardif). MAVS then initiates activation of IRF-3/7 and NF-κB transcription factors and the subsequent production of IFNs and inflammatory cytokines.

Recent studies have revealed that cytosolic DNA recognition involves multiple factors including DAI (DNA-dependent activator of IRFs, also known as DLM-1 or ZBP1) [22], IFI16 (interferon, gamma-inducible protein 16) [23], DDX41 (DEAD box polypeptide 41) [24], and others. IFN-stimulatory DNA (ISD) [25] is a synthetic 45-mer ds-DNA often used as a ligand that can activate a cytosolic DNA pathway, which is different from B-DNA-mediated RIG-I-dependent pathway. Another DNA ligand, dsVACV 70 mer, which was found as an IFN-β-inducible viral ds-DNA that is conserved in various poxviral genomes such as vaccinia virus (VACV) [23], is used to activate IFI16-mediated signaling. Most of these cytosolic DNA sensors induce IFNs and inflammatory cytokines through the adaptor protein stimulator of interferon genes (STING) (Fig. 1) [8]. Most recently, it has been demonstrated that STING itself is an innate immune sensor for cyclic diguanylate monophosphate (cyclic di-GMP), which is a second messenger in bacteria [26]. Furthermore, cyclic GMP-AMP (cGAMP), which is synthesized by host cGAMP synthetase (cGAS) from cytosolic DNA, has been identified as a ligand for STING to

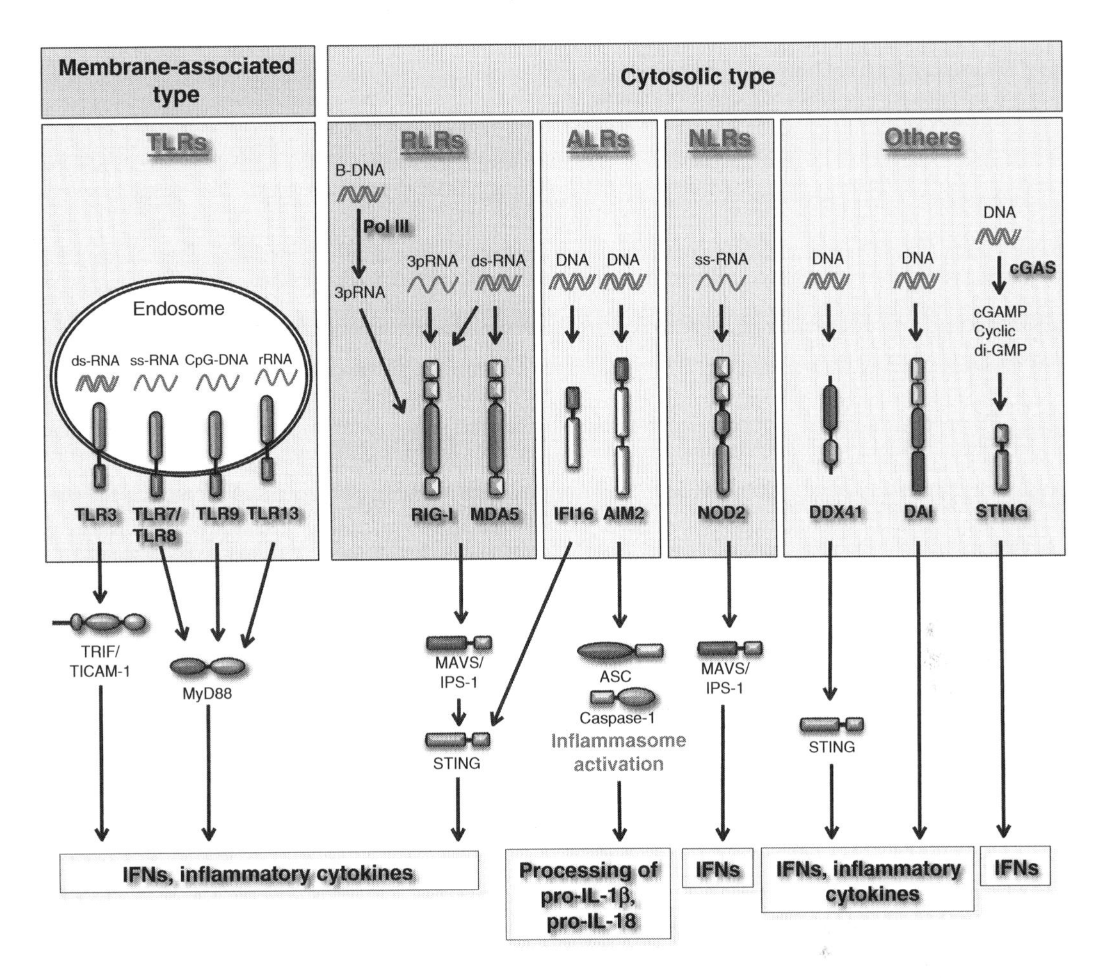

Fig. 1 Nucleic acid-mediated activation of innate immune signalings. Nucleic acid sensors are mainly divided into two groups: Membrane-associated type and cytosolic type. Nucleic acid-ligand binding to sensors results in their association with their adaptor proteins such as TRIF, MyD88, MAVS, or STING, leading to the induction of IFNs and inflammatory cytokines. On the other hand, some types of DNA also activate inflammasome, leading to the maturation of IL-1β and IL-18. Abbreviations: *TLRs* Toll-like receptors, *RLRs* RIG-I-like receptors, *NLRs* NOD-like receptors, *ALRs* AIM2-like receptors, *ds-RNA* double-stranded RNA, *ss-RNA* single-stranded RNA, *rRNA* ribosomal RNA, *Pol III* RNA polymerase III, *3pRNA* 5′-triphosphate RNA, *cGAMP* cyclic GMP-AMP, *IFNs* interferons

induce type I IFN production during infection with DNA viruses such as herpes simplex virus 1 (HSV-1) [27–29].

In most cases of nucleic acid sensing, the downstream signaling leads to type I and III IFN production, which resultantly activates IFN receptor-mediated signalings [1]. Type I IFNs (e.g., IFN-α, β) bind to a receptor complex composed of two subunits, IFNAR-1 and IFNAR-2. On the other hand, type III IFNs (also known as IFN-λs) interact with a different receptor complex composed of IL-28Rα and IL-10R2. Both types of IFNs commonly utilize the JAK-STAT signaling pathway to form the ISGF3 transcription factor complex, which targets IFN-stimulated regulatory element (ISRE) of the promoter region of IFN-stimulated genes (ISGs), such as *OAS1*, *OAS2*, *ISG15*, *Mx*, and *PKR*, to confer an antiviral state on uninfected cells [30].

On the other hand, there is a cytosolic DNA recognition pathway that activates inflammasome to induce inflammatory cytokines but not type I IFNs [4, 31]. AIM2 (absent in melanoma 2), one of the ALR family members, binds to B-DNA and leads to activation of the inflammasome complex together with the adaptor protein ASC (apoptotic speck protein containing a CARD) and caspase-1 [32]. Inflammasome acts as a platform for the activation of the proinflammatory caspase-1, wherein the active form of caspase-1 proteolytically cleaves the cytosolic-sequestering leader sequences from pro-IL-1β, pro-IL-18, and pro-IL-33 to generate their mature form to mediate the downstream inflammatory events during infection with *Listeria monocytogenes*, *Mycobacterium tuberculosis*, *Francisella tularensis*, and DNA viruses, such as adenovirus, murine cytomegalovirus, vaccinia virus, and HSV-1 [4, 31, 33].

In this chapter, we first describe the preparation of various types of RNA/DNA ligands and how to stimulate cells extracellularly or intracellularly with these nucleic acid ligands. Next we explain the experimental procedures of dual-luciferase assay, quantitative RT-PCR (qRT-PCR), and enzyme-linked immunosorbent assay (ELISA) to evaluate the induction of IFNs and proinflammatory cytokines in response to nucleic acid ligands. These methods described here would be useful to characterize a pathophysiological role of microbial or host nucleic acid-mediated innate immune signalings in the development of infectious, autoimmune, and inflammatory diseases.

2 Materials

2.1 Preparation of Ligands for Extracellular RNA/DNA Stimulation

1. RNA ligands: Poly(rI:rC) as a TLR3 ligand is commercially obtainable from GE Healthcare or InvivoGen. Both ss-RNA and R848 for TLR7/8 stimulation are from InvivoGen, and ORN Sa19 as a TLR13 ligand from InvivoGen.
2. DNA ligands: Type-B CpG oligonucleotides (ODNs) for TLR9 stimulation can be purchased from InvivoGen, or its custom-made ODNs are available from Sigma. Each sequence is below with a phosphorothioate backbone, as indicated in lowercase: ODN 2006 (a human TLR9 ligand), 5′-tcgtcgttttgtcgttttgtcgtt-3′; ODN1826 (a mouse TLR9 ligand), 5′-tccatgacgttcctgacgtt -3′; the above ODNs with the replacement of CpG with GpC are often used as a negative control (*see* **Note 1**).

2.2 Preparation of Ligands for Cytosolic RNA/DNA Stimulation

1. RNA ligands: 3pRNA ligand for RIG-I stimulation is prepared by in vitro transcription. Briefly, template DNA (*see* below) is transcribed by T7 RNA polymerase using the MEGAscript T7 Kit (Ambion), and the transcribed RNA is purified by using ISOGEN (Nippon Gene). The sequence of template DNA containing T7

promoter region (indicated by underline) is as follows: 5′-<u>TAATACGACTCACTATAGG</u>GAAACTAAAAGGGAGAAGTGAAAGTG-3′ (*see* **Note 2**). Poly(rI:rC) as RIG-I and/or MDA5 ligands can commercially be obtained from GE Healthcare or InvivoGen (*see* **Note 3**).

2. DNA ligands: B-DNA [poly(dA:dT)·(dT:dA)] is commercially available from Sigma or InvivoGen. For other types of ds-DNA ligands, the following oligonucleotides are synthesized through Sigma, and complementary strands are annealed: dsVACV 70 mer (*see* Subheading 1 for details and [23]); sense, 5′-CCATCAGAAAGAGGTTTAATATTTTTGTGAGACCATCGAAGAGAGAAAGAGATAAAACTTTTTTACGACT-3′; antisense, 5′-AGTCGTAAAAAAGTTTTATCTCTTTCTCTCTTCGATGGTCTCACAAAAATATTAAACCTCTTTCTGATGG-3′. ISD (*see* Subheading 1 for details and [25]); sense, 5′-TACAGATCTACTAGTGATCTATGACTGATCTGTACATGATCTACA-3′; antisense, 5′-TGTAGATCATGTACAGATCAGTCATAGATCACTAGTAGATCTGTA-3′. Cyclic di-GMP, cyclic di-AMP, and cGAMP, all of which are used as STING ligands, are commercially obtainable from InvivoGen.

2.3 Cell Culture

1. Human embryonic kidney (HEK) 293 T cells; THP-1 cells, a human monocytic cell line; RAW 264.7 cells, a murine macrophage cell line. These cells are obtainable from American Type Culture Collection or RIKEN BioResource Center through the National Bio-Resource Project of the MEXT, Japan. For purification of human primary $CD14^+$ monocytes, human peripheral blood mononuclear cells (PBMCs) are prepared from whole blood by density gradient centrifugation using Ficoll PM400 (Sigma). Human primary $CD14^+$ monocytes (>95 % $CD14^+$ as determined by flow cytometry) are obtained from PBMCs by magnetic activated cell sorting with magnetic microbeads according to the manufacturer's instructions (Miltenyi Biotec) (*see* **Note 4**).
2. Culture medium: Dulbecco's modified Eagle's medium (DMEM) supplemented with 10 % heat-inactivated fetal bovine serum (FBS) is used for HEK293T and RAW264.7 cells; RPMI-1640 medium supplemented with 10 % heat-inactivated FBS is for THP-1 cells and primary $CD14^+$ monocytes.
3. 0.2 % Trypsin–2 mM EDTA solution.
4. Phosphate-buffered saline (PBS) solution: 136.9 mM NaCl, 2.7 mM KCl, 10 mM Na_2HPO_4, 2.0 mM KH_2PO_4 (adjust to pH 7.4 with HCl if necessary).
5. Opti-MEM I Reduced-Serum Medium (Invitrogen).
6. Transfection reagents: Lipofectamine2000 (Invitrogen), FuGENE HD (Promega).

2.4 Dual-Luciferase Assay

1. Reporter plasmids: p-125Luc plasmid [34] is used as IFN-β-luciferase reporter. p-55C1BLuc [34], NF-κB-Luc (Clontech or Promega), or pISRE-Luc (Clontech) for IRF-, NF-κB-, or ISRE-luciferase reporter, respectively. pRL-TK (Promega) is used for Renilla luciferase plasmid as an internal control.
2. Dual-Luciferase Reporter (DLR®) Assay System (Promega): Stop & Glo® Buffer, Stop & Glo® Substrate, luciferase assay substrate, luciferase assay buffer II, passive lysis buffer (5×).
3. Luminometer capable of reading 96-well plates: CentroXS LB960 (Berthold).
4. White opaque 96-well plates (Nunc).

2.5 Quantitative RT-PCR

1. RNA isolation: Total RNAs from cultured cells are isolated by using ISOGEN (Nippon Gene).
2. DNase reagent: DNase I, Amplification Grade (Invitrogen).
3. Reverse transcription reagent: ReverTra Ace qPCR RT Kit (TOYOBO).
4. PCR reagent: SYBR Premix Ex Taq (Tli RNaseH Plus) (TAKARA).
5. Oligonucleotide primers: Make 100 μM master stock and 10 μM working stock using nuclease-free water. Primer sequences are shown in Table 2.
6. Thermocycler: StepOnePlus™ real-time PCR system (Applied Biosystems).
7. MicroAmp® Optical 96-well Reaction plate (Applied Biosystems).

2.6 Enzyme-Linked Immunosorbent Assay

1. Human or murine IFN-β: VeriKine™ Human Interferon Beta ELISA Kit (PBL), VeriKine™ Mouse Interferon Beta ELISA Kit (PBL).
2. Human IL-6: Human IL-6 Platinum ELISA (eBioscience).
3. Human IL-1β: IL-1β, Human, ELISA Kit (R&D Systems).

3 Methods

Carry out all procedures at room temperature unless otherwise specified.

3.1 Extracellular RNA/DNA Stimulation

1. Cells are seeded at $1–3 \times 10^5$ cells (HEK293T, THP-1, and RAW264.7 cells) or $5–10 \times 10^5$ cells (primary $CD14^+$ monocytes) in 12-well plate in 500 μl of culture medium with serum but without antibiotics and grown overnight.
2. 0.5–50 μg of RNA/DNA ligands are suspended in 500 μl of culture medium (*see* **Note 5**).

Table 2
Oligonucleotides used for the amplification of human and murine genes

Species	Gene symbol	Forward primer (5′–3′)	Reverse primer (5′–3′)	References
Human	*IFNA1*	GCCTCGCCCTTTGCTTTACT	CTGTGGGTCTCAGGGAGATCA	Nat Immunol 2011, 12:37–44
	IFNA4	ACCTGGTTCAACATGGAAATG	ACCAAGCTTCTTCACACTGCT	Nat Immunol 2011, 12:37–44
	IFNB1	ATGACCAACAAGTGTCTCCTCC	GCTCATGGAAAGAGCTGTAGTG	Nat Immunol 2011, 12:37–44
	CXCL10	GTGGCATTCAAGGAGTACCTC	GCCTTCGATTCTGGATTCAGACA	Nat Immunol 2011, 12:37–44
	TNF	ATGAGCACTGAAAGCATGATCC	GAGGGCTGATTAGAGAGAGGTC	Nat Immunol 2011, 12:37–44
	IL6	AACCTGAACCTTCCAAAGATGG	TCTGGCTTGTTCCTCACTACT	Nat Immunol 2011, 12:37–44
	OAS2	AACTGCTTCCGACAATCAAC	CCTCCTTCTCCCTCCAAAA	J Immunol 2003, 170:749–756
	ISG15	GAGAGGCAGCGAACTCATCT	CTTCAGCTCTGACACCGACA	Breast Cancer Res 2008, 10:R58
	ACTB	CATGTACGTTGCTATCCAGGC	CTCCTTAATGTCACGCACGAT	Nat Immunol 2011, 12:37–44
	GAPDH	CATGAGAAGTATGACAACAGCCT	AGTCCTTCCACGATACCAAAGT	Nat Immunol 2011, 12:37–44
Murine	*Ifna4*	TGATGAGCTACTACTGGTCAGC	GATCTCTTAGCACAAGGATGGC	J Immunol 2010, 185:6146–6156
	Ifnb1	GAGCTCCAAGAAAGGACGAAC	GGCAGTGTAACTCTTCTGCAT	J Immunol 2010, 185:6146–6156
	Cxcl10	CCAAGTGCTGCCGTCATTTTC	GGCTCGCAGGGATGATTTCAA	Nucl Acid Res 2012, 40:D1144–D1149
	Tnf	CCCTCACACTCAGATCATCTTCT	GCTACGACGTGGGCTACAG	Infect Immun 2005, 73:3990–3998
	Il6	TAGTCCTTCCTACCCCAATTTCC	TTGGTCCTTAGCCACTCCTTC	J Immunol 2008, 181:2723–2731
	Oas1	TGTCCTGGGTCATGTTAATAC	CCGTGAAGCAGGTAGAGA	J Immunol 2008, 180:2474–2485
	Isg15	AGCAATGGCCTGGGACCTAAA	AGCCGGCACACCAATCTT	J Virol 2006, 80:4501–4509
	Actb	GCTCCCCGGGCTGTATTCC	CTCTCTTGCTCTGGGCCTCGT	J Virol 2006, 80:4501–4509
	Gapdh	AGGTCGGTGTGAACGGATTTG	TGTAGACCATGTAGTTGAGGTCA	J Immunol 2010, 185:6146–6156

3. Medium is removed from the cells, and the above 500-μl mixture is added gently to the cells.
4. Cells are stimulated for 3 h with the required nucleic acid ligands according to your experimental plan (*see* **Note 6**).

3.2 Cytosolic RNA/DNA Stimulation

1. Prepare cells seeded as described in Subheading 3.1.
2. 0.5 μg of RNA/DNA ligands are suspended in 125 μl of Opti-MEM and 1 μl of Lipofectamine2000 in 125 μl of Opti-MEM (*see* **Note 7**). Each suspension is placed at room temperature for 5 min and then gently mixed with each other. The mixture is incubated for 20 min.
3. After incubation, add 250 μl of culture medium. Replace cell culture medium gently with this 500-μl mixed solution.
4. Cells are stimulated for 6–12 h (for qRT-PCR) or 24 h (for dual-luciferase assay and ELISA) post-transfection with the required nucleic acid ligands according to your experimental plan (*see* **Note 6**).

3.3 Dual-Luciferase Assay

1. For 24-well plate transfections, cells are seeded at $1–3 \times 10^5$ cells (HEK293T and RAW264.7 cells) per well and grown overnight.
2. Cells are transfected with 100 ng (HEK293T cells) or 500 ng (RAW264.7 cells) of luciferase reporter plasmids of interest, together with 10 ng (HEK293T cells) or 50 ng (RAW264.7 cells) Renilla luciferase reporter plasmid using Lipofectamine2000 (HEK293T) or FuGENE HD (RAW264.7 cells) transfection reagent according to the manufacturer's instruction (*see* **Note 8**).
3. Cells are then stimulated for 24 h with nucleic acid ligands as described in Subheading 3.1 or 3.2.
4. After medium is removed from the cells, the cells are washed with PBS and lysed with 100 μl of 1× Passive Lysis Buffer (Promega). Luminescence is measured with the luminometer CentroXS LB960.
5. Firefly luminescence activity is normalized for transfection efficiency by dividing with Renilla activity.

3.4 qRT-PCR

1. Cells are stimulated for 6–12 h with nucleic acid ligands as described in Subheading 3.1 or 3.2.
2. After stimulation, the cells are washed with PBS and then added with 500 μl of ISOGEN for RNA isolation.
3. Total RNAs are isolated according to the manufacturer's instruction, and the RNA pellet is resuspended in nuclease-free water.
4. 1 μg of RNA is treated with DNase I and reverse-transcribed according to the manufacturer's instruction.

5. For the detection of mRNA expression levels of IFNs and other cytokines, each primer set indicated in Table 2 is used for PCR analysis with SYBR Premix Ex Taq reagent and StepOnePlus real-time PCR system. Specifically, the PCR reaction mixture is assembled in 0.2-ml 96-well plates in the following order by adding 10 μl of SYBR Premix Ex Taq II (Tli RNaseH Plus) (2×), 0.4 μl of 10 μM specific forward and reverse primers, 0.4 μl of ROX Reference Dye (50×), and 7.8 μl of nuclease-free water. Finally, 1 μl of the reverse-transcribed cDNA is added to the reaction mixture.
6. The samples are subjected to the PCR analysis; a 10-s initial denature step at 95 °C, followed by 45 cycles of the two-step program; denaturation at 98 °C for 5 s; and combined primer annealing/extension at 60 °C for 30 s.
7. Data are normalized to the expression of *ACTB/Actb* or *GAPDH/Gapdh* for each sample.

3.5 ELISA

1. Cells are stimulated with nucleic acid ligands as described in Subheading 3.1 or 3.2.
2. After incubation at 37 °C in a humidified atmosphere of 5 % CO_2, the samples are centrifuged at 500 × *g* for 3 min, and the supernatant is then transferred to a new tube.
3. The cytokine concentration in each sample is measured according to the manufacturer's instruction.

4 Notes

1. TLR9 ligand ODNs containing an unmethylated CpG motif are classified into three groups: types A, B, and C. Each ODN has a distinct effect on the innate immune response. Type A CpG ODNs, which are characterized by poly G tails with phosphorothioate linkages flanking a central palindromic CpG motif-containing sequence with a phosphodiester backbone, induce high levels of type I IFNs in plasmacytoid dendritic cells but have a small effect on the NF-κB activation. On the other hand, both type B and C CpG ODNs have a fully phosphorothioate backbone and lack poly G tails. Type B CpG ODNs strongly activate B cells but weakly produce type I IFNs. Type C CpG ODNs have both features of types A and B. Representative type A and C CpG ODNs are shown below with a phosphorothioate backbone as indicated in lowercase and a palindromic region as indicated by underline: ODN2216 (type A, a human TLR9 ligand), 5′-ggGGGACGATCGTCgggggg-3′; ODN1585 (type A, a mouse TLR9 ligand), 5′-ggGGTCAACGTTGAgggggg-3′; and ODN M362 (type C, a human and mouse TLR9 ligand), 5′-tcgtcgtcgttcgaacgacgttgat-3′ (*see* ref. 35).

2. T7 RNA polymerase-transcribed RNAs have a triphosphate modification at the 5′-end and its self-complementarity at the 3′-end, which is caused by the elongation in a second step through the RNA-dependent RNA polymerase activity of T7, the so-called copy-back mechanisms (*see* ref. 36). 3pRNA is also commercially available from InvivoGen.
3. The length of poly(rI:rC) is a determinant for the activation of RIG-I or MDA5. In general, long poly(rI:rC) (>~2 kb) is recognized by MDA5, and shorter poly(rI:rC) is recognized predominantly by RIG-I (*see* ref. 17, 18).
4. Proposals for research using human blood cells should undergo a process of ethical review and approval.
5. For RNA stimulation, poly(rI:rC), R848 (imidazoquinoline compound), or ORN Sa19 (a 19-mer oligoribonucleotide derived from *Staphylococcus aureus* 23S rRNA, 5′-GGACGGA AAGACCCCGUGG-3′; *see* ref. 10) is used at the final concentration of 10–100 μg/ml, 0.01–10 μg/ml, or 0.02–2 μg/ml, respectively. For TLR9 stimulation, each type of CpG ODN is used at the final concentration of 1–5 μM or 1–50 μg/ml.
6. Duration of stimulation should be optimally determined in the case of each type of ligands and cells tested. In general, cells are stimulated with nucleic acid ligands for 6–24 h. For qRT-RCR analysis, the induction of *IFNB1/Ifnb1* mRNA can be detected after 3–6 h post-stimulation with cytosolic nucleic acids. A preliminary time course analysis is recommended to determine the optimal time duration for ligand stimulation. For dual-luciferase reporter assay and ELISA, the cells are usually harvested 24–48 h after ligand stimulation.
7. Upon stimulation with cytosolic RNA/DNA, nucleic acid ligands are used at the final concentration of 1–10 μg/ml. For STING activation, cells are stimulated with cyclic di-GMP, cyclic di-AMP, or cGAMP according to the methods for extracellular RNA/DNA stimulation (*see* Subheading 3.1) with slight modification by digitonin permeabilization. Briefly, cells are incubated for 30 min at 37 °C with these STING ligands in 500 μl of permeabilization buffer (50 mM HEPES, pH 7, 100 mM KCl, 3 mM $MgCl_2$, 0.1 mM DTT, 85 mM sucrose, 0.2 % BSA, 1 mM ATP, and 0.1 mM GTP) with 10 μg/ml digitonin (Sigma). Permeabilization buffer is then replaced with culture medium with 10 % FBS and incubated for several hours according to your experimental plan (*see* ref. 26). These STING ligands are used at the final concentration of 1–100 μg/ml.
8. Transfection condition should be optimized for each type of cell. The appropriate choice of a transfection reagent is required for efficient transfection. If transfection efficiency is too low, the amount of each plasmid vector can be increased up to 1.0 μg. Ratios of 10:1–50:1 (or greater) for experimental vector:co-reporter vector combinations are feasible.

References

1. Takaoka A, Yanai H (2006) Interferon signalling network in innate defence. Cell Microbiol 8:907–922
2. Mogensen TH (2009) Pathogen recognition and inflammatory signaling in innate immune defenses. Clin Microbiol Rev 22:240–273
3. Takeuchi O, Akira S (2010) Pattern recognition receptors and inflammation. Cell 140: 805–820
4. Vladimer GI, Marty-Roix R, Ghosh S, Weng D, Lien E (2013) Inflammasomes and host defenses against bacterial infections. Curr Opin Microbiol 16:23–31
5. Kono H, Rock KL (2008) How dying cells alert the immune system to danger. Nat Cell Biol 8:279–289
6. Jounai N, Kobiyama K, Takeshita F, Ishii KJ (2012) Recognition of damage-associated molecular patterns related to nucleic acids during inflammation and vaccination. Front Cell Infect Microbiol 2:168. doi:10.3389/fcimb.2012.00168
7. Barrat FJ, Meeker T, Gregorio J, Chan JH, Uematsu S, Akira S et al (2005) Nucleic acids of mammalian origin can act as endogenous ligands for Toll-like receptors and may promote systemic lupus erythematosus. J Exp Med 202:1131–1139
8. Barbalat R, Ewald SE, Mouchess ML, Barton GM (2011) Nucleic acid recognition by the innate immune system. Annu Rev Immunol 29:185–214
9. Takaoka A, Taniguchi T (2008) Cytosolic DNA recognition for triggering innate immune responses. Adv Drug Deliv Rev 60: 847–857
10. Oldenburg M, Krüger A, Ferstl R, Kaufmann A, Nees G, Sigmund A et al (2012) TLR13 recognizes bacterial 23S rRNA devoid of erythromycin resistance-forming modification. Science 337:1111–1115
11. Hemmi H, Takeuchi O, Kawai T, Kaisho T, Sato S, Sanjo H et al (2000) A Toll-like receptor recognizes bacterial DNA. Nature 408:740–745
12. Onoguchi K, Yoneyama M, Fujita T (2011) Retinoic acid-inducible gene-I-like receptors. J Interferon Cytokine Res 31:27–31
13. Yoneyama M, Fujita T (2009) RNA recognition and signal transduction by RIG-I-like receptors. Immunol Rev 227:54–65
14. Rehwinkel J, Reis e Sousa C (2010) RIGorous detection: exposing virus through RNA sensing. Science 327:284–286
15. Yoneyama M, Kikuchi M, Natsukawa T, Shinobu N, Imaizumi T, Miyagishi M et al (2004) The RNA helicase RIG-I has an essential function in double-stranded RNA-induced innate antiviral responses. Nat Immunol 5:730–737
16. Fujita T (2009) A nonself RNA pattern: tri-p to panhandle. Immunity 31:4–5
17. Takeuchi O, Akira S (2009) Innate immunity to virus infection. Immunol Rev 227:75–86
18. Kato H, Takeuchi O, Mikamo-Satoh E, Hirai R, Kawai T, Matsushita K et al (2008) Length-dependent recognition of double-stranded ribonucleic acids by retinoic acid-inducible gene-I and melanoma differentiation-associated gene 5. J Exp Med 205:1601–1610
19. Ablasser A, Bauernfeind F, Hartmann G, Latz E, Fitzgerald KA, Hornung V (2009) RIG-I-dependent sensing of poly(dA:dT) through the induction of an RNA polymerase III-transcribed RNA intermediate. Nat Immunol 10:1065–1072
20. Chiu Y-H, MacMillan JB, Chen ZJ (2009) RNA polymerase III detects cytosolic DNA and induces type I interferons through the RIG-I pathway. Cell 138:576–591
21. Sabbah A, Chang TH, Harnack R, Frohlich V, Tominaga K, Dube PH et al (2009) Activation of innate immune antiviral responses by Nod2. Nat Immunol 10:1073–1080
22. Takaoka A, Wang Z, Choi MK, Yanai H, Negishi H, Ban T et al (2007) DAI (DLM-1/ZBP1) is a cytosolic DNA sensor and an activator of innate immune response. Nature 448: 501–505
23. Unterholzner L, Keating SE, Baran M, Horan KA, Jensen SB, Sharma S et al (2010) IFI16 is an innate immune sensor for intracellular DNA. Nat Immunol 11:997–1004
24. Zhang Z, Yuan B, Bao M, Lu N, Kim T, Liu Y-J (2011) The helicase DDX41 senses intracellular DNA mediated by the adaptor STING in dendritic cells. Nat Immunol 12:959–965
25. Stetson DB, Medzhitov R (2006) Recognition of cytosolic DNA activates an IRF3-dependent innate immune response. Immunity 24:93–103
26. Burdette DL, Monroe KM, Sotelo-Troha K, Iwig JS, Eckert B, Hyodo M et al (2011) STING is a direct innate immune sensor of cyclic di-GMP. Nature 478:515–518
27. Sun L, Wu J, Du F, Chen X, Chen ZJ (2013) Cyclic GMP-AMP synthase is a cytosolic DNA sensor that activates the type I interferon pathway. Science 339:786–791
28. Wu J, Sun L, Chen X, Du F, Shi H, Chen C et al (2013) Cyclic GMP-AMP is an endogenous second messenger in innate immune signaling by cytosolic DNA. Science 339: 826–830

29. Ablasser A, Goldeck M, Cavlar T, Deimling T, Witte G, Röhl I et al (2013) cGAS produces a 2′-5′-linked cyclic dinucleotide second messenger that activates STING. Nature 498(7454): 380–384
30. Sadler AJ, Williams BRG (2008) Interferon-inducible antiviral effectors. Nat Cell Biol 8:559–568
31. Lamkanfi M, Dixit VM (2012) Inflammasomes and their roles in health and disease. Annu Rev Cell Dev Biol 28:137–161
32. Hornung V, Ablasser A, Charrel-Dennis M, Bauernfeind F, Horvath G, Caffrey DR et al (2009) AIM2 recognizes cytosolic dsDNA and forms a caspase-1-activating inflammasome with ASC. Nature 458:514–518
33. Rathinam VAK, Vanaja SK, Fitzgerald KA (2012) Regulation of inflammasome signaling. Nat Immunol 13:333–342
34. Yoneyama M, Suhara W, Fukuhara Y, Sato M, Ozato K, Fujita T (1996) Autocrine amplification of type I interferon gene expression mediated by interferon stimulated gene factor 3 (ISGF3). J Biochem 120:160–169
35. Krieg AM (2002) CpG motifs in bacterial DNA and their immune effects. Annu Rev Immunol 20:709–760
36. Schmidt A, Schwerd T, Hamm W, Hellmuth JC, Cui S, Wenzel M et al (2009) 5′-triphosphate RNA requires base-paired structures to activate antiviral signaling via RIG-I. Proc Natl Acad Sci U S A 106:12067–12072

Chapter 4

Principles for the Use of In Vivo Transgene Techniques: Overview and an Introductory Practical Guide for the Selection of Tetracycline-Controlled Transgenic Mice

Norio Sakai

Abstract

Transgenic mice are a beneficial tool that can allow researchers to investigate the roles of specific genes in physiology and disease. However, conventional transgenic mice have the limitation that constitutive expression of a transgene from the embryonic stage may affect the normal development of the mice or cause compensating effects. To overcome these disadvantages, tetracycline-controlled transgenic mice, which can express target gene products in a tissue-specific and time-dependent manner, have been developed. In this section, the principles of tetracycline-controlled systems are discussed first. In addition, useful information for generating transgenic mice using this system is introduced.

Key words Tetracycline-controlled system, Transgenic mice, Tet-off system, Tet-on system, Gene silencing

1 Introduction

Transgenic mice have provided beneficial information about the functions of specific genes in vivo. In conventional transgenic mice, gene products of interest are constitutively expressed under the control of tissue-specific promoters. However, this constitutive system provides only limited information about the real functions of the target gene. For example, if the promoters used are active during the embryonic stage, the expressed gene products are likely to influence development. There are always questions remaining about whether compensatory effects occurred in response to the presence of the transgene product during embryogenesis. For this reason, constitutive overexpression systems are imperfect tools for analyzing gene function.

To address this problem, inducible overexpression transgenic systems have been developed. Among several inducible systems that are currently available, the most widely used system is a

Shunichi Shiozawa (ed.), *Arthritis Research: Methods and Protocols*, Methods in Molecular Biology, vol. 1142, DOI 10.1007/978-1-4939-0404-4_4, © Springer Science+Business Media New York 2014

tetracycline-controlled transcriptional activation system. In this chapter, the benefits of this system and hints for applying this system to your research are outlined.

2 Materials

In this section, tetracycline-controlled transgenic mice and plasmid vectors, which are used for expressing gene products of interest in a tissue-specific and time-dependent manner, are introduced. Detailed information is described in the below Subheading 3.

3 Methods

3.1 Principles of the Tetracycline-Controlled Transcriptional Activation Systems

Among tetracycline-controlled transcriptional activation systems, two commonly used systems are Tet-off and Tet-on. The tetracycline-controlled transcriptional silencer (tTS) system is also used in combination with the Tet-on system for tight regulation of target protein expression. In addition, the tTS system is utilized for tetracycline-controlled inducible knockdown of target genes.

3.1.1 Tet-Off System (Fig. 1)

In the Tet-off system, a tetracycline-controlled transactivator (tTA), which was developed by Gossen et al., is used for transcriptional activation (*see* **Note 1**). To make Tet-off-controlled transgenic mice, two lines of transgenic mice are necessary. One line must have a transgene encoding tTA under the control of a promoter that provides tissue- or cell-specific expression. The other line should have a transgene encoding the target gene under the control of the minimal promoter sequence from human cytomegalovirus promoter IE (PCMV) combined with the tet operator sequences from *E. coli* (TetO). In a bitransgenic mouse that was made by crossing these two lines, tTA will be produced in a tissue- or a cell type-specific manner. In the absence of tetracycline (or its derivative, doxycycline—Dox), tTA will bind to TetO and activate PCMV, which can induce the transcription of the target gene. In the presence of Dox, tTA undergoes a conformational change and will no longer bind to TetO, turning off the transcription of target genes. The Tet-off system can strongly induce genes when it is "on," while the leakage of transcripts is very low in the presence of Dox (when it is "off").

3.1.2 Tet-On System (Fig. 2)

In the Tet-off system, Dox must be continuously administered when the transcription of a target gene needs to be silenced during embryonic and neonatal stages. This treatment may result in undesirable effects. In addition, it takes several days to obtain the target gene product after the cessation of Dox administration, with the exact

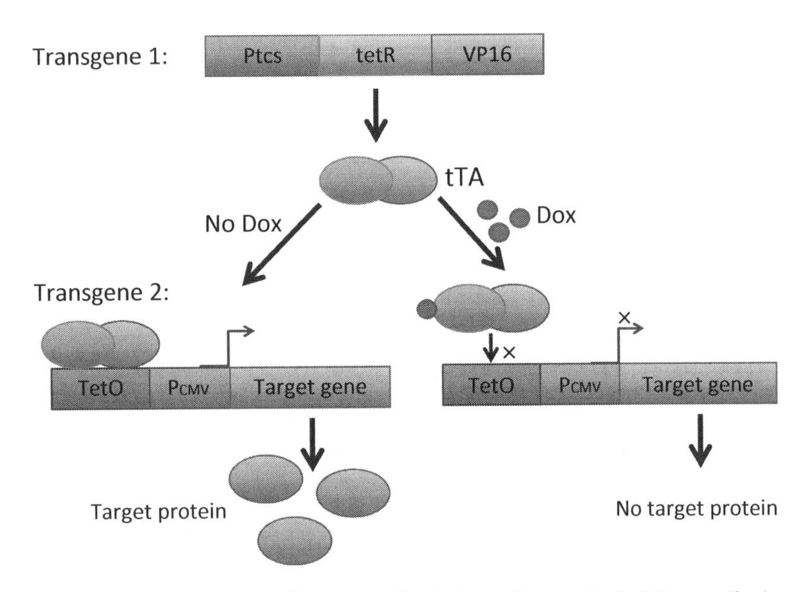

Fig. 1 Illustration of Tet-off system. The tetracycline-controlled transactivator (tTA), which is generated by fusing the DNA-binding domain of the tetracycline-resistance operon (TetR) encoded in *Escherichia coli* Tn10 with the transcription activation domain of the herpes simplex virus virion protein 16 (VP16) [1], is produced under the control of a tissue- or a cell type-specific promoter (Ptcs) from transgene 1. In the absence of doxycycline (Dox), tTA binds to the tetracycline operator sequences of *E. coli* (TetO) and activates the CMV promoter (PCMV), which induces transcription of the target gene. In the presence of Dox, tTA undergoes a conformational change and cannot bind to TetO, switching off transcription of the target gene

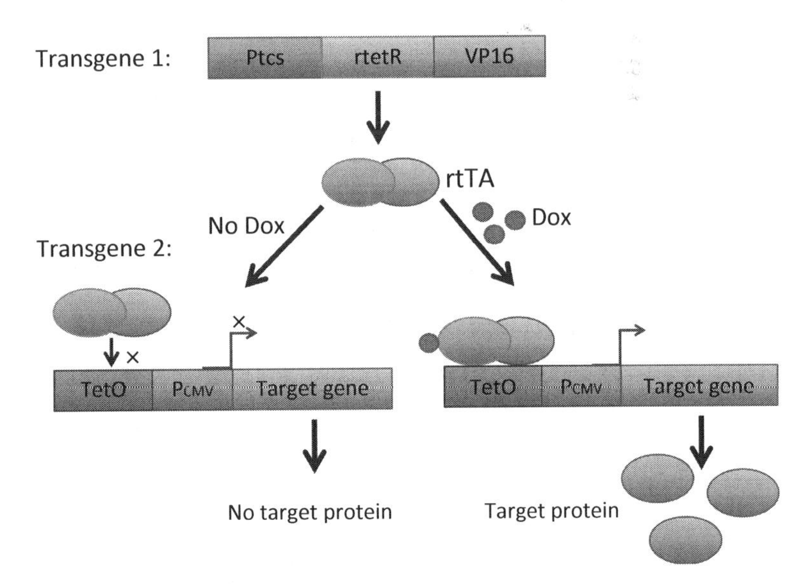

Fig. 2 Illustration of Tet-on system. The reverse tetracycline-controlled transcriptional activator (rtTA) binds to tetO in the presence of Dox. In the absence of Dox, rtTA cannot bind to TetO. The transactivation of the target gene occurs in the presence of Dox, rather than in the absence of Dox

time depending on the rate at which Dox is cleared from the system. To overcome this disadvantage, the Tet-on inducible system was developed by Gossen et al. (*see* **Note 2**). In this system, the reverse tetracycline-controlled transcriptional activator (rtTA) is used for transcriptional activation. In contrast to tTA, rtTA binds to TetO in the presence of Dox. In the absence of Dox, rtTA is no longer able to bind to TetO. Therefore, the transactivation of the target gene occurs in the presence of Dox rather than in the absence of Dox. In general, Dox induces the target gene rapidly, within hours. The rtTA system has been successfully applied in a variety of transgenic mice. To date, more advanced versions of rtTA have been developed and available (*see* **Notes 3** and **4**).

3.1.3 Tetracycline-Controlled Transcriptional Silencer Systems (Fig. 3)

One of the disadvantages of the Tet-on system is that rtTA has the ability to bind to TetO with low affinity even in the absence of Dox, resulting in the leaky induction of the target gene. To overcome this problem, a tTS system has been generated (*see* **Note 5**). Like tTA, tTS can bind to TetO sequences in the absence of Dox; however, tTS cannot activate PCMV because tTS has a silencing domain. In the presence of Dox, tTS undergoes a conformational change that causes it to dissociate from the TetO sequences. In transgenic mice that have the tTS system in combination with rtTA, rtTA can bind to TetO and activate transcription of the target gene in the presence of Dox. In contrast, tTS cannot bind to TetO in the presence of Dox, so it does not affect the rtTA-induced transcription of the target gene. In the absence of Dox, tTS binds to TetO and suppresses the transcription of the target gene, preventing leaky induction of target gene by rtTA. The tTS system is effective for the tight regulation of target genes in the Tet-on system.

3.1.4 Tetracycline-Controlled Inducible Gene Silencing Systems Using tTS (Fig. 4)

To make transgenic mice with a tetracycline-controlled gene silencing system, it is necessary to prepare a transgene that can express short hairpin RNA (shRNA) under the control of a tetracycline-responsive promoter derived from the TetO sequence and the human U6 promoter (TetO/U6) (*see* **Note 6**). In bitransgenic mice that carry both transgenes, a specific promoter-tTS and a TetO/U6-shRNA, tTS can control the expression of the shRNA in a tetracycline-inducible manner. In the absence of Dox, tTS binds to TetO/U6 and suppresses the expression of shRNA, preventing silencing of the gene corresponding to the shRNA sequences. In the presence of Dox, tTA undergoes a conformational change and is unable to bind to TetO/U6, resulting in the expression of the shRNA and silencing the gene corresponding to the shRNA sequences.

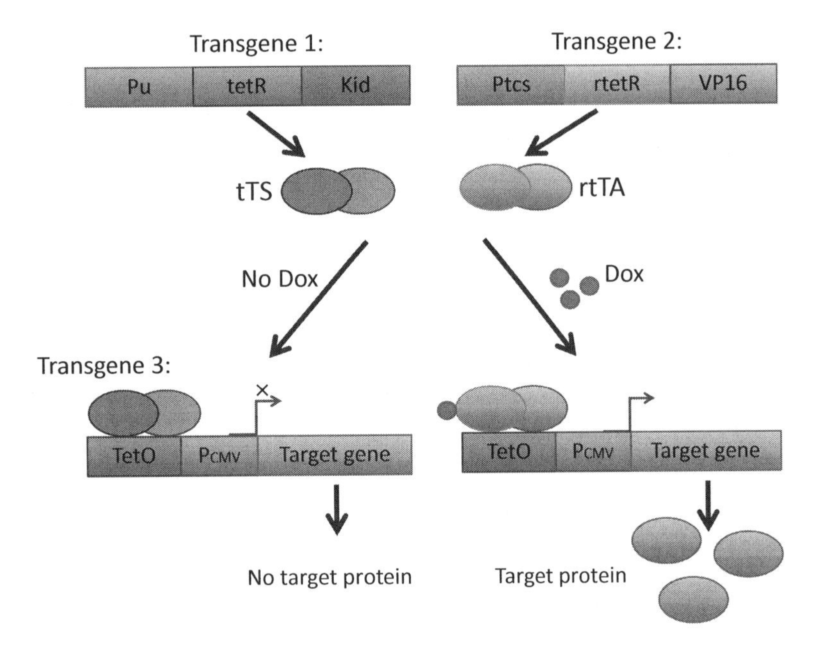

Fig. 3 Illustration of tTS system. The tetracycline-controlled transcriptional silencer (tTS) is made by fusing TetR with the KRAB-AB domain of the Kid-1 protein [4], a transcriptional repressor. tTS is expressed under the control of a universal promoter (Pu). In tritransgenic mice that have the tTS system in combination with rtTA, rtTA can bind to TetO and activate transcription of the target gene, but tTS cannot bind to TetO in the presence of Dox; therefore, tTS does not affect the rtTA-induced transcription of the target gene in the presence of Dox. In the absence of Dox, tTS binds to TetO and tightly suppresses the transcription of the target gene

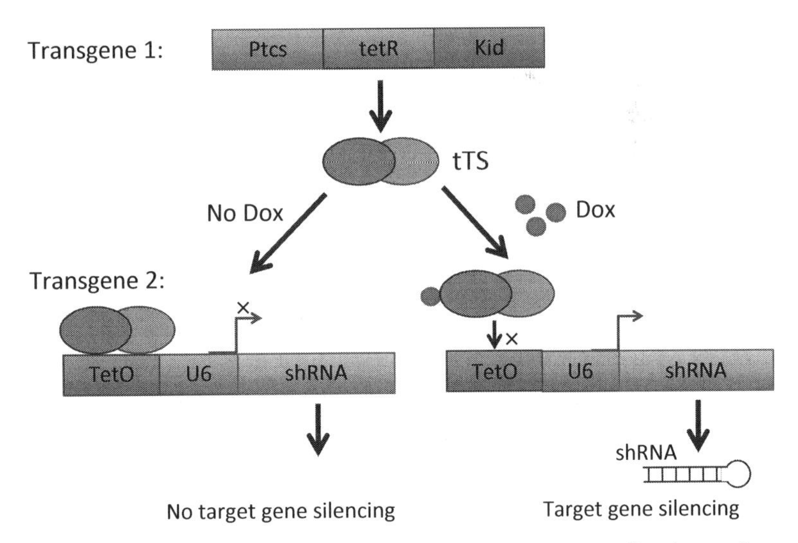

Fig. 4 Illustration of a tetracycline-controlled inducible gene silencing system. In the absence of Dox, tTS binds to the TetO/U6 promoter and suppresses the expression of the shRNA, preventing silencing of the gene corresponding to the shRNA sequences. In the presence of Dox, tTS undergoes a conformational change and cannot bind to TetO/U6, resulting in the expression of the shRNA and the silencing of the gene corresponding to the shRNA target sequences

3.2 Applications of Tetracycline-Controlled Transgenic Mice Suitable for Experimental Purposes

3.2.1 Use of Preexisting Tetracycline-Controlled Transgenic Mice

The Jackson Laboratory provides a variety of commercially available tetracycline-controlled mice. Purchasing from this source is the easiest way to acquire tetracycline-controlled transgenic mice to further research interests. On their Web site, one can search for suitable tetracycline-controlled transgenic mice (*see* **Note 7**). Alternatively, a PubMed search reveals that more than 360 articles reporting the use of such mice have been published to date, and suitable mice may be found in these articles in the field of immunology and rheumatology (*see* **Note 8**).

3.2.2 Generation of Original Transgenes That Can Express Target Proteins Under Tetracycline-Controlled Systems

Clontech provides a variety of plasmid vectors that are helpful in producing original transgenes that induce the expression of tTA or rtTA in a tissue- or a cell type-specific manner (*see* **Note 9**). In these plasmids, CMV promoters are located upstream of the tTA or the rtTA. An original transgene, which can drive the tTA or the rtTA expression in the desired tissue or cell, can be obtained by replacing these CMV promoters with the desired tissue- or cell type-specific promoter. Clontech also provides vectors that are useful for producing transgenes that can induce the expression of target proteins in a tetracycline-regulated manner (*see* **Note 10**). In addition, a suitable plasmid vector for making transgenic mice that can express tTS in a tissue-specific manner can be purchased from Clontech (*see* **Note 11**). Several lines of plasmid vectors with tetracycline-controlled devices are also available as a T-REx system from Life Technologies.

3.2.3 Generation of Tetracycline-Controlled Transgenic Mice by Crossing Suitable Lines of Mice

In principle, two lines of mice, tissue-specific promoter-tTA or rtTA mice and TetO-target gene mice, can be crossed with each other to produce the desired tetracycline-controlled mice. The tetracycline-controlled system can function only in bitransgenic mice, which have both transgenes. If the tTS system is also used in tetracycline-controlled transgenic mice, tritransgenic mice are required. Therefore, it takes a significant investment of time to make tetracycline-controlled mice. However, if stable transgenic mouse lines that can tightly induce the tissue-specific expression of tTA or rtTA are obtained once, these lines can be repeatedly crossed with a variety of TetO-target gene mice, resulting in various transgenic mice with tissue-specific expression of a variety of target proteins. For this reason, tetracycline-controlled transgenic mice are suitable for use in making many types of transgenic mice with tissue-specific expression of target genes.

4 Notes

1. A tTA was generated by fusing the DNA-binding domain of the tetracycline-resistance operon (TetR) encoded in *Escherichia coli* Tn10 with the transcription activation domain of the herpes simplex virus virion protein 16 (VP16) (*see* ref. 1).

2. The rtTA was formed by fusing mutant TetR with VP16 (*see* ref. 2).
3. More advanced versions of rtTA, such as rtTA-S2 or rt-M2, which have higher sensitivity to Dox, were developed by Urlinger et al. (*see* ref. 3).
4. Recently, a plasmid vector containing a more advanced version of rtTA (Tet-on 3G tetracycline-inducible expression systems) has become available from Clontech: http://www.clontech.com/JP/Products/Inducible_Systems/Tetracycline-Inducible_Expression/Tet-On_3G?sitex=10020:22372:US#.
5. A tTS system was made by fusing TetR with the KRAB-AB domain of the Kid-1 protein, which is a transcriptional repressor (*see* ref. 4).
6. For the construction of tetracycline-controlled gene-silencing mice, plasmid vectors such as pSIREN-RetroQ-TetH or pSIREN-RetroQ-TetP are available from Clontech: http://www.clontech.com/US/Products/Inducible_Systems/Tetracycline-Inducible_Expression/shRNA.
7. The Web site of the Jackson Laboratory (http://jaxmice.jax.org/index.html) is a resource for finding suitable tetracycline-controlled transgenic mice. At present, more than 110 mice with Tet-off systems and 50 mice with Tet-on systems are available.
8. For example, CD34-tTA mice (*see* Ref. 5), MMTV–tTA mice (*see* Ref. 6), SCL-tTA mice (*see* Ref. 7), CD4-tTA mice (*see* Refs. 8 and 9), and CD68-rtTA mice (*see* Ref. 10) are available to induce the cell type-specific expression of target proteins in immune cells. The Col I-2.3-tTA mice can be used for osteoblast-specific expression (*see* Ref. 11).
9. Typical plasmid vectors for this purpose include pCMV-Tet3G, pTet-On-Advanced, and pTet-Off-Advanced, as shown below: http://www.clontech.com/JP/Products/Inducible_Systems/Tetracycline-Inducible_Expression/Selection_Guide?sitex=10025:22372:US.
10. pTRE-Tight and pTRE3G vectors are useful for this purpose. In these vectors, the gene of interest can be inserted under the tetracycline-responsive promoter. pTRE3G is currently the most advanced version, in which the tetracycline-controlled system is tightly regulated. See reference Web sites below.
    - pTRE-Tight: http://www.clontech.com/xxclt_ibcGetAttachment.jsp?cItemId=17960.

- pTRE3G:
  http://www.clontech.com/JP/Products/Inducible_Systems/Tetracycline-Inducible_Expression/ibcGetAttachment.jsp?cItemId=44969&fileId=6080382&sitex=10025:22372:US.
  http://www.clontech.com/JP/Products/Inducible_Systems/Tetracycline-Inducible_Expression/Gen2?sitex=10025:22372:US.

11. The pTet-tTS vector is suitable for making transgenic mice that can express tTS in a tissue-specific manner:
http://www.clontech.com/JP/Products/Inducible_Systems/Tetracycline-Inducible_Expression/ibcGetAttachment.jsp?cItemId=17944&fileId=5878229&sitex=10025:22372:US.

References

1. Gossen M, Bujard H (1992) Tight control of gene expression in mammalian cells by tetracycline-responsive promoters. Proc Natl Acad Sci U S A 89:5547–5551
2. Gossen M, Freundlieb S, Bender G, Muller G, Hillen W, Bujard H (1995) Transcriptional activation by tetracyclines in mammalian cells. Science 268:1766–1769
3. Urlinger S, Baron U, Thellmann M, Hasan MT, Bujard H, Hillen W (2000) Exploring the sequence space for tetracycline-dependent transcriptional activators: novel mutations yield expanded range and sensitivity. Proc Natl Acad Sci U S A 97:7963–7968
4. Freundlieb S, Schirra-Muller C, Bujard H (1999) A tetracycline controlled activation/repression system with increased potential for gene transfer into mammalian cells. J Gene Med 1:4–12
5. Huettner CS, Koschmieder S, Iwasaki H, Iwasaki-Arai J, Radomska HS, Akashi K et al (2003) Inducible expression of BCR/ABL using human CD34 regulatory elements results in a megakaryocytic myeloproliferative syndrome. Blood 102:3363–3370
6. Kharas MG, Yusuf I, Scarfone VM, Yang VW, Segre JA, Huettner CS et al (2007) KLF4 suppresses transformation of pre-B cells by ABL oncogenes. Blood 109:747–755
7. Bockamp E, Antunes C, Maringer M, Heck R, Presser K, Beilke S et al (2006) Tetracycline-controlled transgenic targeting from the SCL locus directs conditional expression to erythrocytes, megakaryocytes, granulocytes, and c-kit-expressing lineage-negative hematopoietic cells. Blood 108:1533–1541
8. Huai J, Firat E, Niedermann G (2007) Inducible gene expression with the Tet-on system in CD4+ T cells and thymocytes of mice. Genesis 45:427–431
9. Rahim MM, Chrobak P, Hu C, Hanna Z, Jolicoeur P (2009) Adult AIDS-like disease in a novel inducible human immunodeficiency virus type 1 Nef transgenic mouse model: CD4+ T-cell activation is Nef dependent and can occur in the absence of lymphopenia. J Virol 83:11830–11846
10. Pillai MM, Hayes B, Torok-Storb B (2009) Inducible transgenes under the control of the hCD68 promoter identifies mouse macrophages with a distribution that differs from the F4/80 – and CSF-1R-expressing populations. Exp Hematol 37:1387–1392
11. Peng J, Bencsik M, Louie A, Lu W, Millard S, Nguyen P et al (2008) Conditional expression of a Gi-coupled receptor in osteoblasts results in trabecular osteopenia. Endocrinology 149: 1329–1337

Chapter 5

Unraveling Autoimmunity with the Adoptive Transfer of T Cells from TCR-Transgenic Mice

Huard Bertrand

Abstract

Transgenesis of rearranged α and β chains from the T-cell receptor has allowed the generation of a variety of mice with a predetermined T-cell repertoire. These mice have been extensively used as tools to circumvent the low precursor frequency of naturally occurring endogenous T cells. As such, they have been valuable to study pathways of T-cell development in the thymus. In addition, these mice can also be considered as a valuable source of naive and/or memory T cells with a defined specificity. I will comment in this chapter the use of this source of T cells with known antigen reactivity to study in vivo T-cell behavior in the periphery, including during autoimmune responses.

Key words In vivo immunology, T cells, Tolerance, Autoimmunity

1 Introduction

For decades, in vivo mouse T-cell studies in autoimmunity have been limited to end-stage organ destruction, and early stages of T-cell activation such as recognition of the specific antigen were black boxes. We will see in this chapter that T-cell receptor (TCR)-transgenic mice may be used to track in vivo antigen presentation as well as to early qualify the efficiency of the T-cell response induced. Indeed, it is now possible to precisely determine if T cells sense an antigen, where they recognize this antigen in the periphery. Notably, it is also possible to evaluate the quality of the T-cell response induced by being able to discriminate between either an efficient T-cell priming or an abortive response. Adoptive T-cell transfer has shown that T cells react in lymph nodes draining the tissue expressing the specific antigen, even though no end-stage responses are detected. This has been a great surprise since many past investigations concluded that lack of detectable end-stage T-cell responses was due to antigen ignorance [1]. The most common assessment of antigen ignorance is an in vivo assay measuring T-cell proliferation.

Shunichi Shiozawa (ed.), *Arthritis Research: Methods and Protocols*, Methods in Molecular Biology, vol. 1142, DOI 10.1007/978-1-4939-0404-4_5,

Paris et al. established this in vivo assay in 1994 [2]. This revolutionary assay relies on the use of a cell-permeable dye, an ester of carboxyfluorescein diacetate succinimidyl (CFDA-SE), which is cleaved by esterases once entered in the cell cytoplasm to covalently react with amine groups of intracellular proteins. After staining, the dye amount per cell is dependent on cell division, with one cell division resulting in a split of the total fluorescence equally in daughter cells, hence giving a twofold reduction of the fluorescence intensity. In the absence of T-cell proliferation, one can conclude to ignorance, even though one might also consider surface induction of the early T-cell activation marker CD69. It is in fact with extremely low level of antigen expression that T-cell ignorance prevails [3]. Another extremely important conclusion raised from adoptive transfer of TCR-transgenic T cells is that the outcome of the proliferative response can vary from efficient priming to poor abortive response [4]. Efficient T-cell priming with acquisition of effector function such as cytotoxic activity and IFNγ secretion for $CD8^+$ T cells is induced in a favorable environment, largely dependent on the concomitant activation of $CD4^+$ helper T-cell activation [5] and presence of ligands from innate pathogen sensors such as Toll-like receptors [6]. In their absence, abortive proliferation is seen, ending in an anergic state and even in some cases T-cell death [7–10]. Selected experiments described below allow the fine discrimination between T-cell priming and abortive responses, rendering the adoptive transfer of TCR-transgenic T cells an invaluable in vivo approach to assess immunological responses.

To create TCR-transgenic mouse lines, rearranged TCR were cloned from T-cell hybridomas or T-cell clones. The transgenic mice obtained constitute a great advantage compared to the well-known hybridoma instability and the short longevity of T-cell clones in the mouse system, so that many TCR-transgenic animals were generated, for both helper $CD4^+$ and cytotoxic $CD8^+$ T cells. Specificity for the common model antigens such as chicken ovalbumin, pigeon cytochrome C, hen egg lysosome, and SV40 large T antigen were first studied. Later, transgenic mice for these model antigens under the control of tissue-specific promoters were created. Almost all possible organs are now covered own to the cloning of multiple tissue-specific promoters. The type II collagen promoter is available to drive synovial expression of a model antigen [11]. Finally, TCR specific for self-antigens from different organs have been created, optimizing at its best the experimental approach. As an example, the transgenic mouse expressing a TCR for type II collagen is a valuable experimental tool in rheumatoid arthritis [12]. Availability of all these transgenic mice is making possible detailed in vivo investigations covering multiple immunological aspects of immune tolerance and autoimmunity.

2 Materials

TCR-transgenic mice specific for the antigen under investigation.

Immunodominant peptide(s) from the antigen in consideration.

Recipient syngeneic mice.

CFDA-SE (commonly called CFSE, Molecular Probes, Eugene, OR).

Monensin, DMSO, formaldehyde, saponin (Sigma, St. Louis, MO).

Culture medium with 10 % fetal calf serum (FCS).

Fluorescent antibodies against mouse CD4, CD8, cytokines, and CD45 alleles.

Flow cytometer.

3 Methods

Prepare TCR-transgenic T cells from physically dissociated lymph nodes and spleen without any further purification (*see* **Notes 1** and **2**). If desired, purification of TCR-transgenic T cells can be performed prior to injection, but this is not an absolute requirement in most cases.

3.1 Proliferation Assay

1. Dissolve CFSE in DMSO at 10 mM (90 µl DMSO for 500 µg CFSE), aliquot, and store at −20 °C.
2. Wash the T-cell suspension once in PBS, and adjust it to 10×10^6 cells/ml in a 15–50 ml canonical tube.
3. Dilute the CFSE dye 1/4,000 in PBS (do not filter the dye), and use this solution to dilute the T-cell suspension by twofold (final CFSE concentration: 1.25 µM).
4. Incubate at 37 °C for 15 min (shaking is not required).
5. Wash the cells once in PBS, saturate the remaining reactive CFSE by resuspending T cells in culture medium containing 10 % FCS at 10×10^6 cells/ml, and incubate for 30 min at 37 °C.
6. Wash the cells once in PBS, and inject intravenously 10×10^6 cells per mouse.
7. After 3 days, collect and dissociate the organ to study (*see* **Note 3**), and stain the cell suspension with a fluorescent anti-CD4 or anti-CD8 antibody to track transferred TCR-transgenic T cells for 30 min at 4 °C in the dark.
8. Wash once in PBS, and analyze by flow cytometry (*see* **Note 4**). *See* Fig. 1 for a relevant in vivo T-cell proliferation assay.
9. Proliferation can be quantified by assessing the number of divided cells according to the formula $P=[\Sigma(n_i/2_i)\Sigma n_i]\times 100$ and the total number of daughter cells according to the formula

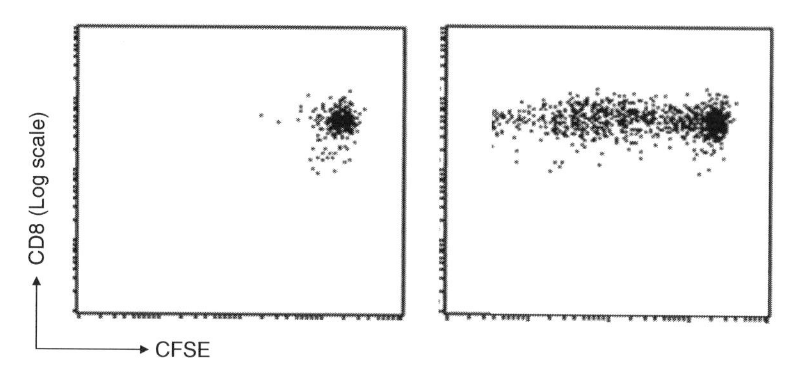

Fig. 1 Proliferation assay. CD8+ OT-I T cells from mice transgenic for an ovalbumin-specific TCR were used. CFSE-labeled OTI-T cells were injected into naive (*left panel*) and ovalbumin peptide-primed (*right panel*) syngeneic C57BL/6 mice. Three days later, control lymph nodes or lymph nodes draining the injection site for the primed mouse were harvested and stained with an anti-CD8 antibody. *Dot plots* show CFSE and CD8 staining for gated $CD8^+$ $CFSE^+$ cells

$N = \Sigma n_i$, n_i being the absolute number of cells for each division rank i.

Full details for quantification can be found in Wells et al. [13].

3.2 Cytokine Secretion Assay

1. Inject intravenously unlabeled TCR-transgenic T cells in recipient animals (*see* **Note 5**).
2. After a minimum of 3 days, collect and dissociate the organ to study, adjust to 10×10^6 cells/ml in culture medium containing 10 % FCS, separate the cell suspension into two, and stimulate one population with a control-irrelevant peptide and the other with the relevant peptide recognized by the transgenic TCR (1 μM) in the presence of the secretion blocker monensin (2 μM) at 37 °C.
3. After 5 h wash cells once in PBS, resuspend at 10×10^6 cells/ml, and add 1 volume of ice-cold 2 % formaldehyde (FA) in PBS while shaking.
4. Incubate for 10 min at room temperature while shaking.
5. Wash once with PBS, and eliminate the remaining reactive FA by washing once in culture medium containing 10 % FCS.
6. Stain cells with a fluorescent allele-specific anti-CD45 and anti-cytokine in buffer containing 1 % saponin for 30 min at RT in the dark.
7. Wash once in PBS 1 % saponin and once in PBS, and analyze intracellular cytokine production in transferred T cells by flow cytometry. *See* Fig. 2 for a relevant cytokine secretion assay.

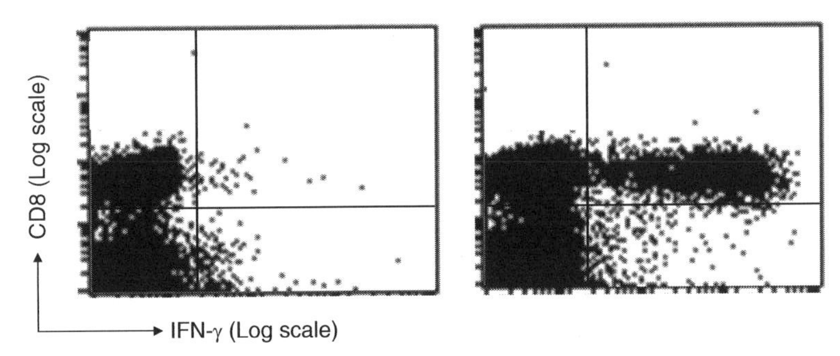

Fig. 2 Cytokine assay. CD8+ OTI-T cells were injected into naive (*left panel*) and ovalbumin peptide-primed (*right panel*) syngeneic C57BL/6 mice. Three days later, control lymph nodes or lymph nodes draining the injection site for primed mice were harvested and restimulated with control (*left panel*) and ovalbumin (*right panel*) peptides, before being stained with anti-CD8 and anti-IFNγ antibodies after cell permeabilization. *Dot plots* show IFNγ and CD8 staining

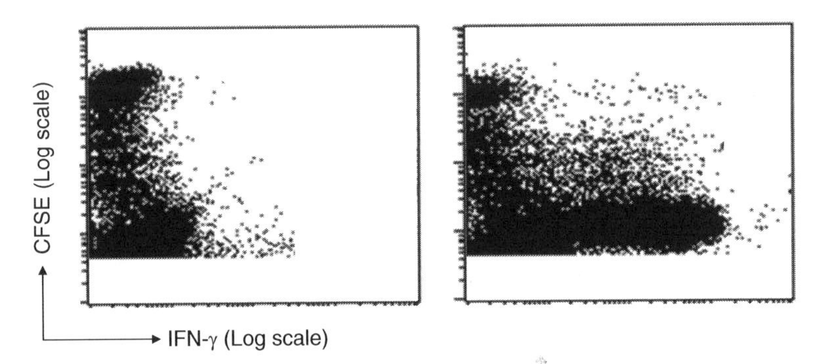

Fig. 3 Proliferation and cytokine assays combined. CFSE-labeled CD8+ OTI-T cells were injected into naive (*left panel*) and ovalbumin peptide-primed (*right panel*) syngeneic C57BL/6 mice. Three days later, control lymph nodes or lymph nodes draining the injection site for primed mice were harvested and restimulated with control (*left panel*) and ovalbumin peptide (*right panel*), before being stained with anti-CD8 and anti-IFNγ antibodies after cell permeabilization. *Dot plots* show IFNγ and CFSE staining on gated CD8+ cells. CFSE- cells were removed from the analysis

8. Note that proliferation and cytokine secretion assays can be combined in a unique experiment. In this case, CFSE-labeled TCR-transgenic T cells are injected into recipient mice. *See* Fig. 3 for a relevant combined assay.

3.3 Cytotoxicity Assay

1. Inject intravenously unlabeled TCR-transgenic T cells in recipient animals (*see* **Note 6**).
2. After 7–9 days collect target cells from the spleen of syngeneic mice. After red blood cell lysis, wash the cell suspension once in PBS, and resuspend it at 10×10^6 cells/ml in PBS.

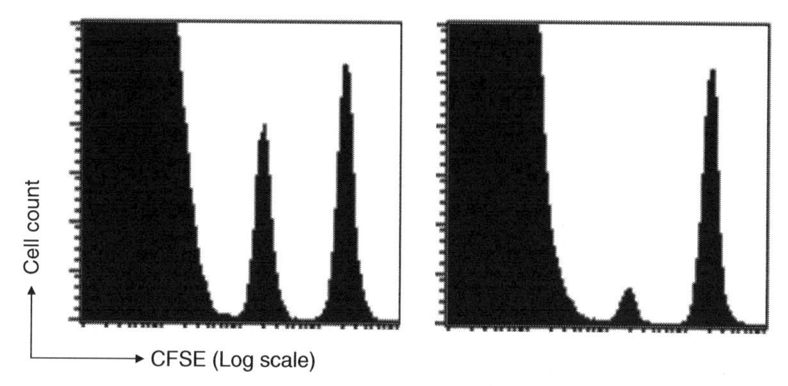

Fig. 4 Cytotoxicity assay. Naive (*left panel*) and ovalbumin peptide-primed (*right panel*) C57BL/6 mice were used. Seven days after priming, control lymph nodes or lymph nodes draining the injection site for primed mice were harvested and analyzed by flow cytometry. *Histogram plots* show the number of dull (ovalbumin peptide-primed) and bright (control peptide-primed) CFSE+ cells recovered

3. Separate the cell suspension into two. Pulse one population with 1 μM of the relevant peptide recognized by the transgenic TCR and the other one with 1 μM of a control-irrelevant peptide for 30 min at 37 °C.
4. Wash once in PBS, and label the cells with CFSE as in Subheading 2.1 to obtain a final CFSE concentration of 0.5 and 5 μM for target cells pulsed with control and relevant peptides, respectively (*see* **Note 7**).
5. Wash the cells once in PBS, saturate the remaining reactive CFSE by resuspending target cells in culture medium containing 10 % FCS at 10×10^6 cells/ml, and incubate for 30 min at 37 °C.
6. Wash the cells once in PBS, count cells, readjust cell concentration if necessary, and inject intravenously 10×10^6 target cells per mouse.
7. After 18 h, collect the organ to study, dissociate, numerate the remaining target cells by flow cytometry, and calculate the ratio $R = \%$ of control target/% of specific target. The percentage of specific cytotoxicity is counted according to the formula $[1 - (R_{\text{naive mouse}}/R_{\text{primed mouse}})] \times 100$. *See* Fig. 4 for a relevant in vivo cytotoxicity assay.

4 Notes

1. Lymph nodes are usually preferred to spleen since they do not require red blood cell lysis.
2. The use of TCR-transgenic mouse on a *RAG*-deficient background avoids sample contamination by non-transgenic T cells due to the leakiness of the allelic exclusion process.

3. Lymph nodes and spleen are the most common organs studied due to the fact that transferred T cells mostly colonize secondary lymphoid organs.
4. The CFSE fluorescent dye is analyzed in the Fl-1 channel and plotted on a logarithmic scale.
5. Use of donor mice expressing a different CD45 allele allows a convenient tracking of transferred T cells in this experiment.
6. Transfer of an exaggerate number of specific T cells has been shown to alter the course of T-cell responses [14], and one seeks to approximate the endogenous frequency of specific T cells upon transfer. While the proliferative and cytokine assays require the detection of transferred cells by flow cytometry implying sufficient number of cells to be transferred, the cytotoxicity assay can be performed with as low as 5×10^4 TCR-transgenic T cells. Such a low number of transferred T cells is in the range of most endogenous T-cell pool for any given specificity.
7. Cytotoxicity against an additional epitope can be studied by adding a third target cell population with an intermediate CFSE staining. We recommend investigators to test the optimal intermediate concentration of CFSE to use since it is dependent on the cytofluorimeter used.

Acknowledgements

Our experience in adoptive T-cell transfer assays has been possible with the financial support from the Swiss National Science Foundation, the Swiss Cancer League, and the Geneva Cancer League.

References

1. Ochsenbein AF, Klenerman P, Karrer U, Ludewig B, Pericin M, Hengartner H et al (1999) Immune surveillance against a solid tumor fails because of immunological ignorance. Proc Natl Acad Sci U S A 96(5): 2233–2238
2. Lyons AB, Parish CR (1994) Determination of lymphocyte division by flow cytometry. J Immunol Methods 171(1):131–137
3. Kurts C, Sutherland RM, Davey G, Li M, Lew AM, Blanas E et al (1999) CD8 T cell ignorance or tolerance to islet antigens depends on antigen dose. Proc Natl Acad Sci U S A 96(22):12703–12707
4. Hernandez J, Aung S, Marquardt K, Sherman LA (2002) Uncoupling of proliferative potential and gain of effector function by CD8(+) T cells responding to self-antigens. J Exp Med 196(3):323–333
5. Kurts C, Carbone FR, Barnden M, Blanas E, Allison J, Heath WR et al (1997) CD4+ T cell help impairs CD8+ T cell deletion induced by cross-presentation of self-antigens and favors autoimmunity. J Exp Med 186(12): 2057–2062
6. Schulz O, Diebold SS, Chen M, Naslund TI, Nolte MA, Alexopoulou L et al (2005) Toll-like receptor 3 promotes cross-priming to virus-infected cells. Nature 433(7028): 887–892
7. Kurts C, Heath WR, Kosaka H, Miller JF, Carbone FR (1998) The peripheral deletion of autoreactive CD8+ T cells induced by cross-presentation of self-antigens involves signaling

through CD95 (Fas, Apo-1). J Exp Med 188(2): 415–420

8. Hernandez J, Aung S, Redmond WL, Sherman LA (2001) Phenotypic and functional analysis of CD8(+) T cells undergoing peripheral deletion in response to cross-presentation of self-antigen. J Exp Med 194(6):707–717
9. Redmond WL, Marincek BC, Sherman LA (2005) Distinct requirements for deletion versus anergy during CD8 T cell peripheral tolerance in vivo. J Immunol 174(4):2046–2053
10. Preynat-Seauve O, Contassot E, Schuler P, Piguet V, French LE, Huard B (2007) Extralymphatic tumors prepare draining lymph nodes to invasion via a T-cell cross-tolerance process. Cancer Res 67(10):5009–5016
11. Zhou G, Garofalo S, Mukhopadhyay K, Lefebvre V, Smith CN, Eberspaecher H, de Crommbrugghe B (1995) A 182 bp fragment of the mouse pro alpha1(II) collagen is sufficient to direct chondrocyte expression in transgenic mice. J Cell Sci 108(Pt12):3677–3684
12. Osman GE, Cheunsuk S, Allen SE, Chi E, Liggit HD, Hood LE, Ladiges WC (1998) Expression of a type II collagen-specific TCR transgene accelerates the onset of arthritis in mice. Int Immunol 10(11):1613–1622
13. Wells AD, Gumundsdottir H, Turka LA (1997) Following the fate of individual T cells throughout activation and clonal expansion. Signals from T-cell receptor and CD28 differentially regulate the induction and duration of a proliferative response. J Clin Invest 100(12): 3173–3183
14. Mintern JD, Davey GM, Belz GT, Carbone FR, Heath WR (2002) Cutting edge: precursor frequency affects the helper dependence of cytotoxic T cells. J Immunol 168(3):977–980

Chapter 6

In Vivo Cell Transfer Assay to Detect Autoreactive T Cell Subsets

Yumi Miyazaki and Shunichi Shiozawa

Abstract

Among the methods used in molecular biology, in vitro biochemical assays are more common, whereas in vivo assays, including the use of animal models, are less widely employed. In our studies on systemic lupus erythematosus (SLE), we have identified a novel T cell subtype termed "autoantibody-inducing CD4 T cells" (*ai*CD4 T cell) that is responsible for the development of autoimmunity. In order to identify and isolate these cells, we developed a new technique that involves the transfer of candidate T cell subpopulations into naïve mice and assaying for the development of autoantibodies in the recipient mice.

We have previously described an experimental system in which mice not normally prone to autoimmune diseases can be induced to develop experimental SLE. In this experimental system, autoantibody-inducing CD4 T (*ai*CD4 T) cells are generated via de novo T cell receptor revision at peripheral lymphoid organs. These *ai*CD4 T cells not only induce a variety of autoantibodies but also promote the final differentiation of CD8 T cells into cytotoxic T lymphocytes, resulting in a pathology identical to SLE. We needed to develop a new methodology to isolate this subpopulation of *ai*CD4 T cells. Here we describe an in vivo assay to detect and isolate *ai*CD4 T cells by transferring the candidate cells into naïve recipient mice and monitoring the production of the appropriate antibody or cytokine.

Key words Systemic lupus erythematosus, Autoantibody-inducing CD4 T cell, Autoantibody, Cell transfer assay, Cell sorting, ELISA

1 Introduction

Systemic lupus erythematosus (SLE) is a systemic autoimmune disease characterized by the presence of autoantibodies in serum, including rheumatoid factor (RF), anti-Sm and anti-dsDNA antibodies, as well as tissue injury such as lupus nephritis, both of which contribute significantly to the pathogenesis of the disease [1–4]. Traditional autoimmune disease theory attributes SLE to the action of peripheral autoreactive lymphocytes that are derived either from a small number of autoreactive clones that have escaped negative selection in the thymus or a small number of clones that have been reactivated from tolerance. However, it is difficult to explain how such restricted sets of clones could account for the

Shunichi Shiozawa (ed.), *Arthritis Research: Methods and Protocols*, Methods in Molecular Biology, vol. 1142, DOI 10.1007/978-1-4939-0404-4_6,

more than 140 distinct specificities of T cell receptor (TCR) and autoantibody repertoires found in SLE [5]. Instead, we have discovered that this repertoire is generated via de novo TCR revision from thymus-passed non-autoreactive clones at peripheral lymphoid organs, resulting in the generation of *ai*CD4 T cells, which stimulate B cells to generate large varieties of autoantibodies. In addition, *ai*CD4 T cells promote the final differentiation of CD8 T cells into cytotoxic T lymphocytes (CTL) via antigen cross-presentation, which in turn cause tissue injuries identical to those seen in SLE [6]. Based on these findings, we have proposed the novel "self-organized criticality theory" of autoimmunity, which explains how wide varieties of autoreactivities can be induced in human SLE by a simple mechanism. By this theory, *ai*CD4 T cells are generated following repeated immunization with antigen by novel V(D)J re-recombination, which occurs at periphery lymphoid regions such as the spleen. Therefore, both the identification and isolation of these *ai*CD4 T cells are essential to elucidate the pathogenesis of SLE. We describe here an experimental cell transfer method by which autoreactive *ai*CD4 T cell subsets can be isolated. This technique represents a novel in vivo cell assay that can detect or monitor in vivo immune alterations.

2 Materials

2.1 Induction of aiCD4 T Cells in Mice

1. Phosphate-buffered saline (PBS), pH 7.4: Prepare this buffer using ultrapure water.
2. BALB/c female mice (Japan SLC, Hamamatsu, Japan).
3. Ovalbumin (OVA) (Sigma Chemical Company, St. Louis, MO.), adjusted to 2 mg/mL with PBS.

2.2 Prepare a Single-Cell Suspension of Mouse Splenocytes

1. 35 mm petri dish: 35 mm/Tissue Culture Dish (ASAHI GLASS CO., LTD., Tokyo, Japan).
2. 50 mL conical tube: BD Falcon 50 mL conical centrifuge tube (BD Falcon, Franklin Lakes, NJ).
3. 15 mL conical tube: 15 mL PP Centrifuge Tube (Corning Inc., Corning, NY).
4. Cell strainer: BD Falcon Cell Strainer (100 μm, BD Falcon).
5. Plunger: TERUMO Syringe (TERUMO CORPORATION, Tokyo, Japan).

2.3 Prepare the Buffer and Reagents for Cell Sorting

1. Working solution: PBS containing 1 % fetal bovine serum (FBS) (dilute 100 μL of FBS with 10 mL of PBS). Keep this solution cold (2–8 °C).
2. AutoMACS Running Buffer (*see* **Note 1**), required for autoMACS Pro Separator assay.
3. Reagents for cell isolation (*see* **Note 2**).

3 Methods

1. Mice (8 weeks old) are repeatedly immunized by injection i.p. 12× with 500 μg of OVA every 5 days (*see* **Note 3**).
2. Using mice that exhibit elevated levels of autoantibodies such as RF or anti-dsDNA, remove spleens and prepare a single-cell suspension (*see* **Note 4**).
   (a) Place spleen on 35 mm petri dish half filled with PBS, and mash the spleen using the plunger.
   (b) Pass homogenized spleen cells through a cell strainer (BD Falcon) mounted on a 50 mL conical tube.
   (c) Rinse the cell strainer with 5 mL PBS, and discard used strainers.
   (d) Transfer suspended cells passed through the strainer to a 15 mL conical tube.
   (e) Determine cell number.
3. Isolate whole CD4 T cells using MACS protocol (Miltenyi Biotec).

   We use the CD4$^+$ T cell isolation kit II as shown below (*see* **Notes 5** and **6**).
   (a) Centrifuge cell suspension at 400 × *g* for 5 min.
   (b) Aspirate supernatant completely, and resuspend cell pellet to 40 μL of the working solution per 1×10^7 total cells.
   (c) Add 10 μL of biotin–antibody cocktail (in the CD4$^+$ T cell isolation kit II) to the tube per 1×10^7 total cells.
   (d) Mix well, and incubate in the refrigerator (2–8 °C) for 10 min.
   (e) Add 30 μL of working solution and 20 μL of Anti-Biotin MicroBeads per 1×10^7 total cells.
   (f) Mix well, and incubate in the refrigerator (2–8 °C) for 15 min.
   (g) Wash cells by adding 5–8 mL PBS and centrifuge at 400 × *g* for 5 min.
   (h) Aspirate supernatant completely.
   (i) Resuspend the pellet of up to 1×10^8 cells in 500 μL of working solution, and proceed to magnetic separation with autoMACS Pro Separator.
4. Isolate particular subpopulation of CD4 T cells via specific surface marker.

   We use Anti-Biotin MicroBeads (Miltenyi Biotec) to isolate respective subpopulations as follows (*see* **Notes 5** and **6**):
   (a) Take the total isolated CD4 T cell suspension and centrifuge at 400 × *g* for 5 min.
   (b) Aspirate supernatant completely.

(c) Resuspend cell pellet in 100 μL of PBS per 1×10^7 total cells. This is then incubated with the primary biotin-conjugated antibody, with the time and titer of the antibody varying depending on the manufacturer's recommendations. Typically, this is on ice for 30 min.

(d) Wash cells by adding 5–8 mL PBS and centrifuge at $400 \times g$ for 5 min.

(e) Aspirate supernatant completely.

(f) Resuspend cell pellet to 80 μL working solution and 20 μL of Anti-Biotin MicroBeads to the tube per 1×10^7 total cells.

(g) Mix well, and incubate in the refrigerator (2–8 °C) for 15 min.

(h) Wash cells by adding 5–8 mL of PBS and centrifuge at $400 \times g$ for 5 min.

(i) Aspirate supernatant completely.

(j) Resuspend the pellet of up to 1×10^8 cells in 500 μL working solution, and proceed to magnetic separation with autoMACS Pro Separator (*see* **Note 7**).

(k) Determine cell number in the sorted CD4 T cell subpopulation. Adjust this sorted cell suspension to 1×10^8 cells/mL with PBS.

5. Transfer 250 μL of sorted cell suspension into one naïve mouse (2.5×10^7 cells/mouse) that has the same background as the donor mice by means of intravenous (i.v.) injection.
6. Two weeks after the cell transfer, collect serum samples from the recipient mice and measure autoantibody levels by means of ELISA (*see* **Notes 8** and **9**).

4 Notes

1. AutoMACS Running Buffer is purchased from Miltenyi Biotec GmbH, Bergisch Gladbach, Germany.
2. Dynabeads (Life Technologies Corporation, Carlsbad, CA) and MACS MicroBeads (Miltenyi Biotec) are commonly used. We use MACS MicroBeads, which are included in the $CD4^+$ T cell isolation kit II, and Anti-Biotin MicroBeads to isolate CD4 T cell subpopulations. To isolate labeled cells, we use an autoMACS Pro Separator purchased from Miltenyi Biotec.
3. Animal studies with BALB/c female mice (Japan SLC, Inc., Hamamatsu, Japan) are performed with the approval of the Institution Review Board.

4. For optimal performance it is important to obtain solely a single-cell suspension and to remove cell clumps by passing homogenized spleen cells through the cell strainer (BD Falcon), before magnetic labeling. Red blood cell lysis is unnecessary.
5. Work fast, keep cells cold in the refrigerator (2–8 °C), and use precooled solutions. This will prevent capping of antibodies on the cell surface and nonspecific cell labeling.
6. Recommended incubation temperature is 2–8 °C. Since the temperature is critical in this assay, if you work at 4 °C, the incubation time required will be approximately 15 min, depending upon the antibodies you use. If you work on ice, longer incubation times may be required, probably around 30 min. Higher temperatures and/or longer incubation times may increase also nonspecific cell labeling.
7. AutoMACS Running Buffer should be warmed to room temperature before use.
8. CD4 T cell subpopulations that can induce autoantibodies in naïve recipients are taken as the *ai*CD4 T cell-enriched subpopulation.
9. Sera were assayed for RF (Shibayagi Co., Gunma, Japan), anti-Sm antibody using Sm antigen (Immuno Vision, Springdale, AR), and anti-dsDNA antibody using dsDNA (Worthington Biochemical Co., Lakewood, NJ) after digestion by S1 nuclease (Promega, Madison, WI).

References

1. Fu SM, Deshmukh US, Gaskin F (2011) Pathogenesis of systemic lupus erythematosus revisited 2011: end organ resistance to damage, autoantibody initiation and diversification, and HLA-DR. J Autoimmun 37:104–112
2. Perry D, Sang A, Yin Y et al (2011) Murine models of systemic lupus erythematosus. J Biomed Biotechnol 2011:271694
3. Arshak-Rothstein A (2006) Toll-like receptors in systemic autoimmune disease. Nat Rev Immunol 6:823–835
4. Christensen AR, Kashgarian M, Alexopolou L et al (2005) Toll-like receptor 9 controls anti-DNA autoantibody production in murine lupus. J Exp Med 202:321–331
5. Shiozawa S (2012) Pathogenesis of SLE and *ai*CD4T cell: new insight on autoimmunity. Joint Bone Spine 79:428–430
6. Tsumiyama K, Miyazaki Y, Shiozawa S (2009) Self-organized criticality theory of autoimmunity. PLoS One 4:e8382

Chapter 7

Characterization of MicroRNAs and Their Targets

Ghada Alsaleh and Jacques-Eric Gottenberg

Abstract

MicroRNAs (miRNAs) have emerged as key players in the degradation of target mRNAs. They have been associated with diverse biological processes, and recent studies have demonstrated that miRNAs play a role in inflammatory responses. The identification of miRNA and their corresponding messenger RNA (mRNA) targets can therefore be very helpful. In this chapter, we first overview the field of miRNAs and then show the fundamental techniques for the identification of miRNAs and confirmation of their role on target mRNAs.

Key words MicroRNA, Autoimmune diseases, Rheumatoid arthritis Jeric

1 Introduction

MicroRNAs (miRNAs) are an evolutionarily conserved class of endogenous small noncoding RNAs. They are produced from long precursor molecules by the consecutive action of the RNase III enzymes DROSHA and DICER, before being loaded on an ARGONAUTE protein within the RNA-induced silencing complex (RISC). The mature miRNA acts as a guide for RISC to mediate destabilization and/or translational repression of target messenger RNAs (mRNAs) [1, 2]. Since their discovery, miRNAs have emerged as key players in the regulation of translation and degradation of target mRNAs [1, 3, 4]. They provide an additional posttranscriptional mechanism by which protein expression can be regulated [5–7]. The expression of 10,000 genes, or 30 % of the human genome, could potentially be regulated by miRNAs [8, 9]. The regulation of miRNA expression is itself controlled at various levels such as transcription, processing, or stability [10] and can be influenced by various stress factors including inflammation [11]. In mammals, miRNAs have been associated with diverse biological processes such as cell differentiation, cancer, regulation of insulin secretion, and viral infection [12, 13].

Shunichi Shiozawa (ed.), *Arthritis Research: Methods and Protocols*, Methods in Molecular Biology, vol. 1142, DOI 10.1007/978-1-4939-0404-4_7, © Springer Science+Business Media New York 2014

1.1 Identification of miRNAs

To illustrate the process of demonstrating the involvement of miRNAs, we have taken in Subheading 3 the example of the search of miRNAs regulating TLR2. miRNA which can target the mRNA of this gene of interest must first be determined through a bioinformatic target prediction tool (*see* **Note 1**); all these sites search for target sequences of the mRNA exclusively located in 3′-UTRs (and not in the 5′-UTR or in protein-coding region [14], where miRNAs can also bind to target sequences of mRNAs, although less frequently as in 3′-UTRs). In all these three computational programs, a specific "Gene Symbol" must be entered to get a list of predicted miRNAs that can target a specific mRNA. Inversely, these programs can also determine all the putative mRNA targets of a given miRNA. However, the information provided by bioinformatic target prediction tools has to be confirmed and validated by experimental assays.

1. The first step usually consists in shortening the usually long list of numerous predicted miRNAs by identifying those among this list whose expression is correlated with the target mRNA and protein studied (in case the mechanism of action of the studied miRNA is mRNA degradation) or only with the target protein (in case the mechanism of action is inhibition of translation). The most frequent situation is an inhibition of the target mRNA/protein and an inverse correlation between the level of the studied miRNA and its target. In our example, it means to study which among the miRNAs predicted to target TLR2 have a decreased level when the level of TLR2 increases. Real-time quantitative PCR (qPCR) is the most frequently used technique to study the expression of the miRNA. In situ hybridization can also be used [15].

   However, some miRNAs, which are not included in the list provided by the predicting tools, might play an important regulatory role. Instead of analyzing only the expression of miRNAs involved in the list by qPCR, cDNA microarrays containing nucleotide probes complementary to miRNAs of human origin can also be used.

2. The first step only helps to select the best candidates among those predicted by informatic tools and suggests that a miRNA, usually few miRNAs, can target specific mRNA(s). To increase the confidence on the regulatory role of a miRNA, evidence is needed. Transfection of an antagomir, an antisense oligonucleotide that inhibits the candidate miRNA, has to result in an increased level of the target mRNA, assessed by qPCR, and/or in an increased level of the target protein assessed by Western blot or ELISA. Alternatively, transfection of the analogue of this miR (mimic) has to lead to the decrease of the expression of the target mRNA and/or protein (*see* **Note 2**).

3. After the experiments of transfection with an antagomir and a mimic, the regulatory role of a miRNA is demonstrated. However, this regulation can still be indirect, which means that the miRNA regulates a regulator of the mRNA (a transcription factor, for example) and not directly the mRNA itself. Demonstrating the direct, physical, interaction between a miRNA and its target mRNA requires a reporter gene assay. Thus, a reporter luciferase gene vector into which is inserted the 3′-UTR of the target gene mRNA is constructed. Then this vector is co-transfected with a mimic of the studied miRNA in HEK293 cells (*see* **Note 3**). The measurement of the luciferase activity allows to determine whether a direct interaction between the studied miRNA and the target mRNA exists. If the miRNA directly targets its mRNA, the mimic will bind to the 3′-UTR sequence and induce a reduction of the relative luciferase activity compared to a control vector, in which the inserted 3′-UTR was mutated to reduce its complementary bases with the studied miRNA.

2 Materials

2.1 Cell Culture

1. Human synoviocytes were isolated from synovial tissues from different RA patients at the time of knee joint arthroscopic synovectomy after informed consent was obtained from patients. HEK293 cells were purchased from the American type culture collection (ATCC).
2. Culture medium: Human synoviocytes were maintained in vol/vol of RPMI-1640 (Invitrogen) and Med199 medium (Invitrogen) supplemented with 10 % heat-inactivated FBS, penicillin (100 UI/l), and streptomycin (100 μg/ml). HEK293 was maintained in Dulbecco's modified Eagle's medium (DMEM) supplemented with 10 % heat-inactivated FBS, penicillin (100 UI/L), and streptomycin (100 μg/ml).
3. 0.05 % Trypsin–2 mM EDTA solution.
4. Phosphate-buffered saline (PBS) solution: 136.9 mM NaCl, 2.7 mM KCl, 10 mM Na_2HPO_4, 2.0 mM KH_2PO_4 (adjust to pH 7.4 with HCl if necessary).
5. Opti-MEM I Reduced-Serum Medium (Invitrogen).

2.2 Quantitative RT-PCR

1. RNA isolation: Total RNAs from cultured cells are isolated by using TRIzol® (Sigma).
2. DNase reagent: DNase I, Amplification Grade (Invitrogen).
3. Reverse transcription reagent: iScriptcDNA Synthesis Kit (BioRad).

Table 1
Oligonucleotides used for the amplification of human genes

Genes	Primers
TLR2 human F	5′-GGCCAGCAAATTACCTGTGT-3′
TLR2 human R	5′-CTCCAGCTCCTGGACCATAA-3
GAPDH human F	5′-GGTGAAGGTCGGAGTCAACGGA-3
GAPDH human R	5′-GAGGGATCTCGCTCGCTCCTGGAAGA-3

4. PCR reagent: SensiMix Plus SYBR kit (Quantace, Corbett Life Science).
5. Real-time qPCR analyses for miRNAs: miScript System and the primers (Qiagen).
6. Oligonucleotide primers: Make 100 μM master stock and 10 μM working stock using nuclease-free water. Primer sequences are shown in Table 1.
7. Thermocycler: *Rotor-Gene™ 6000 real-time PCR machine* (*Corbett Life Science®*).

2.3 Transfection Assay

1. Transfection reagents: Human Dermal Fibroblast Nucleofector™ kit (Lonza) for synoviocytes and Lipofectamine2000 (Invitrogen) for HEK293.
2. Mimic: MiRIDIAN® miR-19 mimic and miRIDIAN miRNA mimic negative control (Dharmacon).
3. Nucleofector™-Device: Nucleofector™ 2b Device (Lonza).

2.4 Enzyme-Linked Immunosorbent Assay

1. Human IL-6: Human IL-6 ELISA Kit (R&D Systems).

2.5 Western Blot

1. Lysis buffer: 1 % Triton X-100, 20 mM Tris–HCl pH 8.0, 130 mM NaCl, 10 % glycerol, 1 mM sodium orthovanadate, 2 mM EDTA, 1 mM phenylmethylsulfonylfluoride, and protease inhibitors.
2. Protein electrophoresis and blotting: Criterion Precast System.
3. Blocked buffer: 1 % bovine serum albumin in TBS (20 mM Tris, pH 7.5, 150 mM NaCl).
4. Antibodies: Anti-TLR2 mouse IgG1 monoclonal antibodies (Imgenex, IMG-319, clone 1030A5.138), anti-GAPDH mouse monoclonal antibodies (Millipore, clone 6C5). Horseradish peroxidase-conjugated goat anti-mouse IgG monoclonal antibodies (eBioscience).
5. Chemiluminescence: Super Signal West Femto Maximum Sensitivity Substrate (Pierce).

2.6 Dual-Luciferase Assay

1. Reporter plasmids: psiCHECK-2 (Promega) is used as 3′-UTR sequence of TLR luciferase reporter.
2. Dual-Luciferase Reporter (DLR®) Assay System (Promega): Stop & Glo® Buffer, Stop & Glo® Substrate, luciferase assay substrate, luciferase assay buffer II, passive lysis buffer (5×).
3. Luminometer capable of reading 96-well plates: CentroXS LB960 (Berthold).
4. White opaque 96-well plates (Nunc).

3 Methods

3.1 Stimulation of Cells for Total RNA Extraction

1. For miRNA studies, synoviocytes (10^6 cells) were stimulated with 2 ml of medium alone or medium containing LPS (1 mg/ml) and BLP (1 mg/ml).
2. After a 3- or 6-h incubation period, total RNA was extracted using TRIzol according to the manufacturer's instruction and the RNA pellet was resuspended in nuclease-free water.
3. Total RNA was treated with DNaseI, 1 μl of DNase I for 10 μg of RNA, and then incubated for 30 min at 37 °C and then 10 min at 75 °C.

3.2 Real-Time Quantitative PCR

We performed an RT-qPCR analysis to confirm the downregulation of miR-19b included in a list of predicted several miRNA candidates possibly targeting TLR2 (Fig.1) (*see* **Note 4**).

1. RNA concentrations were adjusted using a NanoDrop instrument to 500 ng of RNA in 10ìl H_2O DEPC per sample.
2. Reverse transcriptase reactions for miRNA were performed using miScript system as follows:
3. Prepare 5μl H_2O DEPC + 4μl miScript RT buffer 5× +1 μl miScript Reverse Transcriptase Mix.
4. Add 10 μl for each sample.
5. Incubate for 60 min at 37 °C and then 5 min at 95 °C.

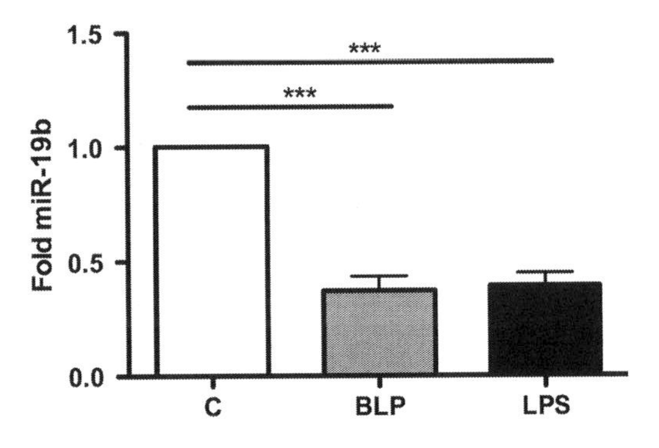

Fig. 1 Downregulation of miR-19b by RT-qPCR

6. For the detection of miRNA expression, each primer was used for PCR analysis with miScript real-time PCR system.
7. Specifically, the PCR reaction mixture was assembled in 20 μl in the following order by adding 10ìl of 2× QuantiTect SYBR Green PCR Master Mix, 1 μl of 10× miScript UP (Universal Primer), 1ìl of miScript Primer Assay (Specific Primer), and 7 μl of RNase-free WATER and 1ìl of the template miRNA.
8. The samples were subjected to PCR and a 15-min initial denature step at 95 °C, followed by 35 cycles of the three-step program: 94 °C for 15 s, 55 °C for 30 s, and 70 °C for 30 s.
9. Expression of endogenous U6 snRNA was used for normalization. Relative expression was calculated using the comparative threshold cycle (Ct) method.

3.3 Transfection Assay or Synoviocytes

1. Prepare cell culture plates by filling appropriate number of wells with 500 μl of culture media, and preincubate plates in a humidified 37 °C/5 % CO_2.
2. Harvest the synoviocyte cells by trypsinization with 2 ml of 0.05 % trypsin–2 mM EDTA solution.
3. Count an aliquot of the cells, and determine cell density.
4. Centrifuge the required number of cells at 300 × *g* for 10 min at room temperature. Remove supernatant completely.
5. Resuspend the cell pellet carefully in room temperature with 100 μl of Nucleofector™ Solution.
6. Add mimics 19b (20 pM/sample) or the mimic miRNA-negative control (20 pM/sample) (*see* **Note 5**).
7. Transfer the mix into the Nucleocuvette™ Vessels.
8. Place Nucleocuvette™ Vessel with closed lid into the retainer of the 4D-Nucleofector™ X Unit, and start the transfection process by pressing the "Start" on the display of the 4D-Nucleofector™ Core Unit.
9. Resuspend cells with 500 μl pre-warmed medium, and mix cells by gently pipetting up and down two to three times.
10. Plate transfected cells in plates described in Subheading 2.
11. After 48 h post-transfection, cells were then stimulated for 6 and 24 h with LPS (1 μg/ml) or BLP (1 μg/ml).
12. Medium was removed from the cells according to the manufacturer's instructions.
13. Culture supernatants were harvested, and IL-6 release was measured by a heterologous two-site sandwich ELISA.
14. The cells were washed with PBS and lysed with 100 μl of ice-cold lysis buffer for protein assay by Western blot.

3.4 Western Blot

We measured TLR2 protein expression by Western blotting in transfected cells by miR-19b mimics or control, and we found

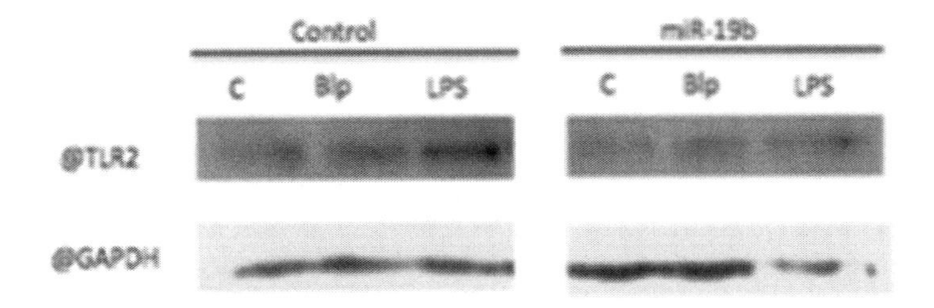

Fig. 2 miR-19b transfection affects Tlr2 protein

that overexpression of miR-19b led to a decrease in TLR2 protein Fig. 2.

1. Cells were transfected and prepared as described in Subheading 2.
2. Lysates were centrifuged for 10 min at 14,000 × g at 4 °C.
3. Supernatants were subjected to SDS-PAGE.
4. Transfer electrophoretically to PVDF membranes.
5. Membranes were blocked using 1 % bovine serum albumin in TBS (20 mM Tris, pH 7.5, 150 mM NaCl) for 1 h at 25 °C.
6. The blots were incubated with anti-TLR2 mouse IgG1 monoclonal antibodies for 2 h at 25 °C.
7. This is followed by incubation with horseradish peroxidase-conjugated goat anti-mouse IgG monoclonal antibodies (1 h at 25 °C).
8. Detect by enhanced chemiluminescence according to the manufacturer's instructions.
9. To confirm the presence of equal amounts of proteins, bound antibodies were removed from the membrane by incubation in 0.2 M glycine, pH 2.8, and 0.5 M NaCl for 10 min at room temperature.
10. Reprobe again with anti-GAPDH mouse monoclonal antibodies.

3.5 Luciferase Reporter Constructs

To verify whether the predicted binding site for miR-19b within Tlr2 mRNA was functional, we generated luciferase reporter constructs.

1. To generate luciferase-based reporter plasmids, psiCHECK-2 (Promega) was modified by inserting the Gateway cassette C.1 (Invitrogen) at the 3′-end of the firefly luciferase gene (f-luc) into the Xba I site of psiCHECK-2.
2. The 3′-UTR sequence of Tlr2 was amplified from HEK293 cell genomic DNA.
3. Add the attB1 and 2 sequences.
4. The resulting PCR products were cloned into pDONR/Zeo and then recombined in the modified psiCHECK-2 vector using Gateway technology (Invitrogen).

3.6 Luciferase Assay

1. We co-transfected the pSI-CHECK2 constructs with an inhibitor of miR-19b (specific antisense oligonucleotides) (*see* **Note 3**).
2. HEK293 cells were plated in 24-well plates (2×10^5 cells per well).
3. Cells were transfected with reporter constructs 50 ng, and miR-19b/a mimic (20 pM/sample) was performed using Lipofectamine2000 (Invitrogen).
4. After 48 h, cells were washed with PBS and lysed with 100 μl of 1× Passive Lysis Buffer (Promega).
5. Firefly luciferase (f-luc) and Renilla luciferase (r-luc) activities were determined using the dual-luciferase reporter assay system (Promega) and a luminometer (Glomax, Promega). The relative reporter activity was obtained by normalization to r-luc activity (Figs. 3 and 4).

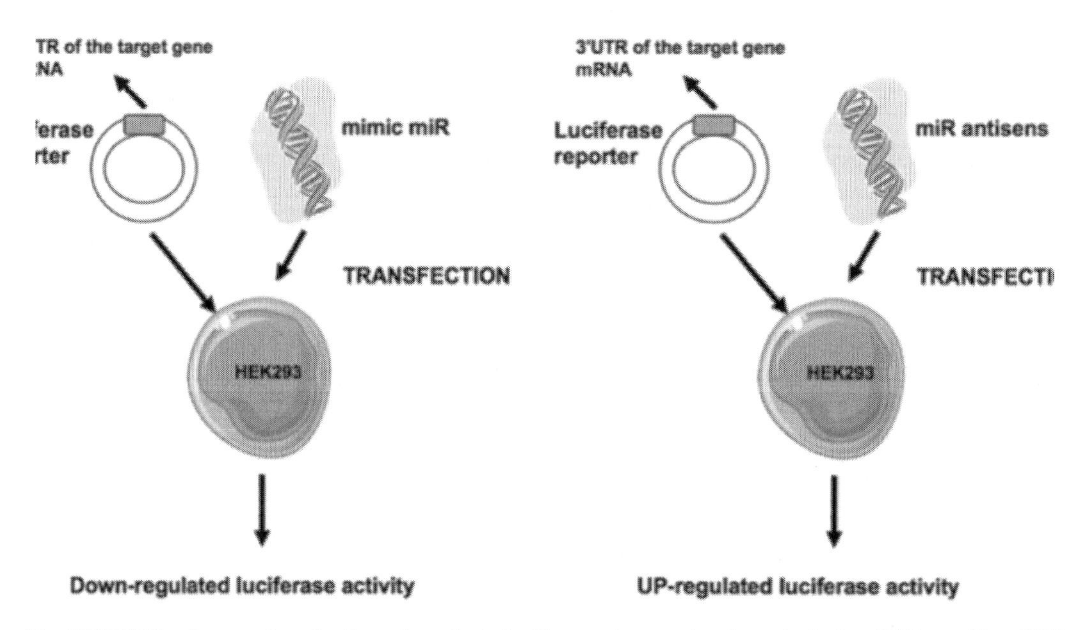

Fig. 3 HEK293 cells were transiently co-transfected with reporter constructs and mimic miRNA or with miRNA antisense molecules. Luciferase activities (RLU) were measured after transfection

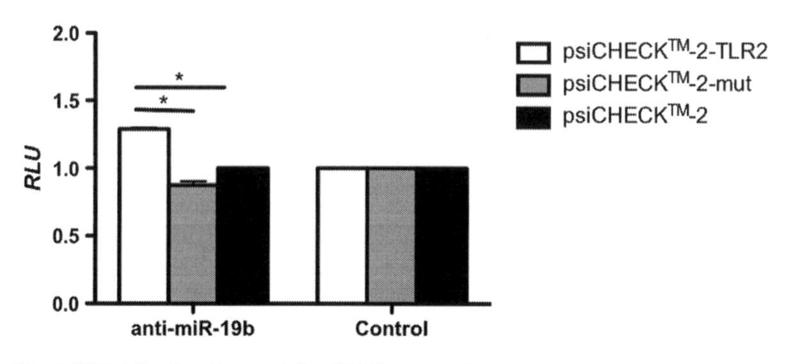

Fig. 4 MiR-19b directly regulates TLR2 expression

4 Notes

1. The three most commonly used prediction tools are:
   (a) Targetscan (http://www.targetscan.org).
   (b) miRanda (http://microrna.sanger.ac.uk).
   (c) PicTar (http://pictar.bio.nyu.edu).
2. We measured TLR2 protein expression by Western blotting in transfected cells by miR-19b mimics or control, and we found that overexpression of miR-19b led to a decrease in TLR2 protein (Fig. 2). In parallel, we measured the secretion of IL-6 by ELISA and found that it was significantly downregulated in BLP-activated FLS transfected with miR-19b mimics.
3. Before testing these constructs in HEK-293 cells, we measured the expression levels of miR-19b in these cells. We found that HEK-293 cells express this miRNA. Therefore we co-transfected the pSI-CHECK2 constructs with control or 19b antisense 2′O methylated oligoribonucleotides instead of mir19b mimic.
4. In the same condition, we determined whether increased TLR2 mRNA expression was correlated with the decreased miR19b expression.
5. Transfection condition should be optimized for each type of cell. We can use concentrations of mimic or inhibitors between 20 and 200 nM.

References

1. Bartel DP (2004) MicroRNAs: genomics, biogenesis, mechanism, and function. Cell 116: 281–297
2. Bartel DP (2009) MicroRNAs: target recognition and regulatory functions. Cell 136: 215–233
3. Griffiths-Jones S, Grocock RJ, van Dongen S et al (2006) miRBase: microRNA sequences, targets and gene nomenclature. Nucleic Acids Res 34:D140–D144
4. Chen K, Rajewsky N (2007) The evolution of gene regulation by transcription factors and microRNAs. Nat Rev Genet 8:93–103
5. Filipowicz W, Bhattacharyya SN, Sonenberg N (2008) Mechanisms of post-transcriptional regulation by microRNAs: are the answers in sight? Nat Rev Genet 9:102–114
6. Pillai RS, Bhattacharyya SN, Filipowicz W (2007) Repression of protein synthesis by miRNAs: how many mechanisms? Trends Cell Biol 17:118–126
7. Lee EJ, Baek M, Gusev Y et al (2008) Systematic evaluation of microRNA processing patterns in tissues, cell lines, and tumors. RNA 14:35–42
8. Lewis BP, Burge CB, Bartel DP (2005) Conserved seed pairing, often flanked by adenosines, indicates that thousands of human genes are microRNA targets. Cell 120: 15–20
9. Rajewsky N (2006) microRNA target predictions in animals. Nat Genet 38 Suppl:S8–S13
10. Krol J, Loedige I, Filipowicz W (2010) The widespread regulation of microRNA biogenesis, function and decay. Nat Rev Genet 11:597–610
11. Leung AKL, Sharp PA (2010) MicroRNA functions in stress responses. Mol Cell 40: 205–215
12. Farazi TA, Juranek SA, Tuschl T (2008) The growing catalog of small RNAs and their association with distinct Argonaute/Piwi family members. Development 135:1201–1214
13. Liang Y, Ridzon D, Wong L et al (2007) Characterization of microRNA expression profiles in normal human tissues. Genomics 8:166
14. Kuhn DE, Martin MM, Feldman DS et al (2008) Experimental validation of miRNA targets. Methods 44:47–54
15. Wheeler G, Valoczi A, Havelda Z et al (2007) In situ detection of animal and plant microRNAs. DNA Cell Biol 26:251–255

Chapter 8

Studies on the T Cell Receptor (TCR) Revision of Autoantibody-Inducing CD4 T (*ai*CD4 T) Cell

Shunichi Shiozawa and Kenichi Uto

Abstract

Our recent studies into the role of autoantibody-inducing CD4 T cells in autoimmune disease have necessitated studies on the mechanism of TCR revision, a phenomenon that has been difficult to approach experimentally. Here we describe a detailed experimental technique to investigate the molecular events involved in TCR revision.

Key words Autoantibody-inducing CD4+ T cell (*ai*CD4+ T cell), T cell receptor (TCR) revision, Recombination-activating gene (RAG), TCR basal signaling

1 Introduction

One of the biggest challenges we face in elucidating the pathogenesis of autoimmunity is understanding how autoreactive lymphocyte clones survive or emerge from the firewall envisioned in the "forbidden clone" theory of Mackay and Burnet [1]. By this theory, peripheral autoreactive lymphocytes are derived either from a few autoreactive clones that have slipped thru the negative selection of the thymus or a few clones that have broken tolerance. However, it is hard to imagine how this relatively small number of "forbidden clones" could account for the wide varieties of TCR and autoantibody repertoires found in systemic lupus erythematosus (SLE), which consist of more than 140 distinct specificities. We have proposed an alternative, novel theory called the "self-organized criticality theory," which postulates the emergence of autoreactive lymphocyte clones by de novo T cell receptor (TCR) revision from non-autoreactive clones at periphery [2]. We have designated this novel T cell type the "autoantibody-inducing CD4

Shunichi Shiozawa (ed.), *Arthritis Research: Methods and Protocols*, Methods in Molecular Biology, vol. 1142, DOI 10.1007/978-1-4939-0404-4_8,

T cell" (*ai*CD4 T cell), and its properties could reasonably account for the multi-specific autoreactivity found in autoimmune diseases such as SLE [2, 3].

Self-organized criticality is a technical term used in systems engineering. The self-organized criticality theory of autoimmunity postulates that systemic autoimmunity, or SLE, is the inevitable consequence of overstimulating the host's immune system by repeated exposure to antigen to levels that surpass the immune system's stability limit, i.e., its self-organized criticality. The *ai*CD4 T cells that emerge as a result stimulate B cells to generate a variety of autoantibodies. They also help to promote the final differentiation of CD8 T cells into cytotoxic T lymphocytes (CTL) via antigen cross-presentation and thus induce lupus tissue injuries [2, 3].

As this *ai*CD4 T cell can cause SLE in mice normally not prone to autoimmune disease, we would like to investigate its properties further. In particular, it is important to afford a CD number on *ai*CD4 T cell and to elucidate the mechanisms by which *ai*CD4 T cells can be generated. To this end, we have extensively studied the TCR Vβ and Jα regions of *ai*CD4 T cells. Here we describe our technique so that the reader may be able to conduct their own investigations and apply these to clinical investigation.

2 Materials

2.1 Induction of aiCD4 T Cell in Mice

1. BALB/c female mice (Japan SLC, Hamamatsu, Japan).
2. Staphylococcus enterotoxin B (SEB) (Toxin Technology, Sarasota, FL).
3. T Cell Isolation Kit II (Miltenyi Biotec, Bergisch Gladbach, Germany).
4. PE-conjugated anti-Vβ8 TCR antibody F23.1 (BD PharMingen, San Diego, CA).
5. Anti-PE microbeads (Miltenyi Biotec, Glandbach, Germany).

2.2 Ligation-Mediated PCR (LM-PCR)

1. BW linker: 5′-GCGGTGACCCGGGAGATCTGAATTC-3′ for BW-1 linker, 5′-GAATTCAGATC-3′ for BW-2 linker, and 5′-CCGGGAGATCTGAATTCCAC-3′ for BW-1H primer (Sigma-Aldrich, St. Louis, MO, USA) [4].
2. Sequences of the locus-specific primers and probes are provided in Tables 1 and 2.
3. Gene Images CDP-Star chemiluminescence reaction (GE Healthcare).

Table 1
The sequences of the locus-specific primers

Sequence (5′–3′)		
Locus	**First primer**	**Second primer**
Vβ2	ggtcaatgtgggtgtgggtaata	tgagctctcagaacttgtcttggt
Vβ4	agtttctcctagcttgactgactgc	acagggctattggtgagtggtaa
Vβ16	gtgtagggggcttttgggtta	gaagtgaagatcaacagggcagt
Vβ10	agatcttggggtgcatagaagtg	acctgaagcctgagtggagaata
Vβ1	gatcctttctcagaagctcagtcc	ggatggcaggacctaaagtatgtg
Vβ5.2	gatagaaccatctgcatgaacacc	ccacacacaaactaagctcagca
Vβ8.3	cctatagaataaacaccggctgga	gggactgaaactatggctgtgag
Vβ5.1	tctatcaccctgccaaacacac	acacacacacaccccgagtt
Vβ8.2	ggggaccttatggccattctat	gggaggggatagggagaaaaa
Vβ8.1	gcacagtacaaatgaggaagcaa	ctcacgaacaaagggactattgact
Vβ13	ctcctgtcagcttatccattttacg	gcacaatataaggatgtcacgaagg
Vβ12	acacacacacacacatggagagag	ggagatgggaccattagtagagga
Vβ11	aggagaggagaagaaaactggtgtc	tcttccatacctcccagcctatt
Vβ9	tcttgcacagcttagacaaaggac	ttggtgactgagatgaaaacgaac
Vβ6	cagaccaaaccagctacaggagt	tttctctctgtctccaaccaacc
Vβ15	ggctttgcaatgaacctgtct	ccgtacttcagcctttcatgg
Vβ20	ttgaggtcacgggagttacaag	acgggagttacaagcagcagtatt
Vβ3	acgatagtgggccagatattcagt	tgatgagcagcagcctttaactt
Vβ7	acagactttccaggtgtttgtgag	aaagagctgaaaggtgtgagtgc
Vβ18	cgatatgaaagctaaggggtgatt	aatttgcccttcatttctcgtg
Vβ14	aggacgcaagcagcagtaaaa	tatcccacttctctctgccaca
5Dβ1	agagaagacgaccatgtagctgtg	accaccgttctaagaagtccagag
5Dβ2	tttacctgggaacgtgaaagagac	atagcccactgcatactacggact
Jα2	ttacaagaactccctctgcctctc	aagacacactggctagagttctggt
Jα5	gctactgcttacaacatggctgac	aacttgcccaagatcacatagctc
Jα6	gaagtaggcatcaggagatccat	ggctctcacgtaatgggaagacta
Jα9	aatcagagaggagggagatacgtg	aacagaagaagagcagagcgtagg
Jα11	gagctaaggaagaaatgatgggaag	cgggtgaaaggtctttatggtc
Jα12	aggagatgcaagcaaaggagtt	aggcagaattcgacttgagtaagg
Jα13	tcaaaataagggtgcagaactgg	tgtgaacaagacaggtcatgtagg

(continued)

Table 1
(continued)

Sequence (5′–3′)		
Locus	**First primer**	**Second primer**
Jα15	cggcagaaagatgatgtagatgtc	cgttgtctgtcagtgaaggtgat
Jα16	tcaatacatgaacgtgcacagc	acgagactgggatgagatcatgt
Jα17	aggcaggcagttgtactagaaagg	tgagaggccctactgactaccatt
Jα18	tcttttagaccgggaaacaggac	actctggcggtggaaagactatt
Jα21	gtattctgacagacgagacggatg	agaattccctgagctagcttgtgt
Jα22	atggcccttgcatactatccttac	taacttgatgccagccagctc
Jα23	tctcaccttcccttcatcttaacc	gacaatatccagagccattaagca
Jα24	gaacacagaagaaggcccaatatc	aaaaggagatcagagaggggtgtt
Jα26	gggactgtttatttaggggaaagc	ccttcctaatgggtttttgaacag
Jα27	atgtgttccctgaggcattgta	aggcattgtactcgggcttta
Jα28	ggggaggggtatagatcaggagta	cagggaaatgcagcaatcaa
Jα30	ggttggtgagcatttcttggtagt	gctagtgacggctataaaggatgc
Jα31	atctttacacccgaccctttctg	cacactgaaaaggctccacactaa
Jα32	agtaagaaagaaggcactggctga	gtcagccaccaactctcttgttta
Jα33	cttcctagcactgcattcattctg	atgcctctagctgttgtgtccat
Jα34	agactgcccagggacctaatttat	acttgaacagtcgcatcttaccc
Jα35	gcctctgtggtctagtgtttctca	tgagaactttacccaggaggaaga
Jα37	gaagcactgactcttgggaaagtt	acctctttctttcccaccacact
Jα38	acgtgtgtgattgtgggaacat	tttatctctgacccttcggaccta
Jα39	tgagttgtgtgctgggagaagt	caaggtggagaagtgtgtgtgac
Jα40	gtcaaagcacccagaataaagagg	cctgcactatcactgactgtttcc
Jα42	gttgctgtgtttgccttacttgag	ctgtgaaaaccttgcctgtgaa
Jα43	agtttaatgtggagagacgggcta	ccctgccatgtcttgatatttacc
Jα45	ggtcttttctgtgccttttcgt	cagcatcatgttgtggttgttg
Jα48	ttaggcatggacactgacagaaag	ttgggcagtgttaggagagtcttt
Jα49	agatggagagaatgacagacatgg	tctggattagcctagggagaatga
Jα50	tgttagctctctggcagttgtcat	acaacatattgaccagagggaagg
Jα52	gtacccaccatttccctctcct	cgagaagggagctaaggagagag
Jα53	catcgtggtttgcctactcct	gaaagctcccaacctaccatcata
Jα56	accaaagtctacatcacagcagtctc	cgttgaccccacatgctattta
Jα57	gaaggaagaaaacaggaaggaagg	cactggtgctttgctctctgtag
Jα58	caatggagggaaacaacactagc	agaagctagggaatcagggagaat

Table 2
The sequences of the locus-specific probes

Locus	Sequence (5′–3′)
Vβ2	ctgaattccacagtggtaaactctgcaggcgcattgaaacaaaaa
Vβ4	ctgaattccacagccctgcagagctcttgcctccctgtacacaaa
Vβ16	ctgaattccacagccttgaaaaaacaccgtcttcctgtacacaaa
Vβ10	ctgaattccacagttgtgcagagtcactgtttccctgtgcacaaa
Vβ1	ctgaattccacagccctgcagactcattagatctctgtacacaaa
Vβ5.2	ctgaattccacagccttacaaagctactggctttctgtaacttaa
Vβ8.3	ctgaattccacagtgatgtgtggcttcctcccctttgcacagaaa
Vβ5.1	ctgaattccacagccttacagagctactggctttctgtaacttaa
Vβ8.2	ctgaattccacagtgatgtggggtttcctcccctctgcacagaaa
Vβ8.1	ctgaattccacagtgatgtgtggcttccttcactctgcacagaaa
Vβ13	ctgaattccacagtaactcagagacacatccctgctgtgcacaaa
Vβ12	ctgaattccacagagctatacaaggatgccctgcctgtgcaaaaa
Vβ11	attccacagagctacagaatcatagtctgctgaggcagaaaccat
Vβ9	ctgaattccacagggctacaatatcacctaccctctgttcataaa
Vβ6	ctgaattccacagtggagcacagctacccccctctccatgcataaa
Vβ15	ctgaattccacagtgctggacaagaacaacacacttcagcaagaa
Vβ20	ctgaattccacagcgctgaagtgtgcactcccccctgcacacaaac
Vβ3	ctgaattccacagcattgaaatgtgcattctctccagtgcataaa
Vβ7	ctgaattccacagctctacacagccacatcctttctacacaaaaa
Vβ18	ctgaattccacagtgctggttgcaagggagaaatctcagcgagaa
Vβ14	tttttgcacagatgtctgccccaccctactcagtgtggaattcag
5Dβ1	gggagggtcctttttttgtataaagctgtaacattgtggaattcag
5Dβ2	ggg aagaaacttt tttgtatcacgatgtaacattgtggaattcag
Jα2	agatgcctaggcttctgtaaaggtgtcacctgcagtggaattcag
Jα5	ctgagggcaggtatttgtactgtgctgtactagggtggaattcag
Jα6	caggagataggtttcatcaaaggctttcctcactgtggaattcag
Jα9	ggaaacagcccattttgtcacaggacaaatcactgtggaattcag
Jα11	gctgcaggccatttttgtggagaggtttgctgctgtggaattcag
Jα12	acctgtgggtgtttttgactgactaagaaacactgtggaattcag
Jα13	ggaaagaagccattttgtaaagacctcacttacagtggaattcag

(continued)

Table 2 (continued)

Locus	Sequence (5′–3′)
Jα15	agccagtgaagaatttgcagggcctcgtttcactgtggaattcag
Jα16	gcaccatttagtttttgtggtggaagagatcactgtggaattcag
Jα17	tagtccttgtgtttttgcttggcttcagatcactgtggaattcag
Jα18	gtaggagcagatttgtgtaaagggggctggcactgtggaattcag
Jα21	aagaagagggctt tctgtaatgg tgctaaccat tgtggaattcag
Jα22	ctctgcagaggttcttgttgttgagcaaatcacagtggaattcag
Jα23	gattggatgtgtttttgacagggtatgtaacacagtggaattcag
Jα24	ctggggacgccattttgtagacgtgtttgtcacagtggaattcag
Jα26	gctctagtgagtttttgctaagaggaaacccactgtggaattcag
Jα27	tccagtcaaggttattgcaatagcacggagcactgtggaattcag
Jα28	catctgtgaagtttttgcaaagaaaggaaattctgtggaattcag
Jα30	gccactagacgttttgggtatggtcccaatcacagtggaattcag
Jα31	cttgcaaagggtttcagtaaaggcaagagatgatgtggaattcag
Jα32	cttcctcagagttattgtaaggctctgcagggctgtggaattcag
Jα33	gacagtgaatgttttgttaaggtttttgtgtctgtggaattcag
Jα34	ggggccaggggtttttgtagaacatgatatcactgtggaattcag
Jα35	ctctgaggaagtttttgttgtagagtcagccactgtggaattcag
Jα37	cctcttgacagtttttgtaaagtgcagcattggggtggaattcag
Jα38	tggctgctccatttctgtaaagctctctataactgtggaattcag
Jα39	tttctaggaggtttttgctgagctggagatcactgtggaattcag
Jα40	ccttaccttggtttatgtagagacacagaacactgtggaattcag
Jα42	ctgcctgctggttattgtaaggccccaaatgactgtggaattcag
Jα43	acaacaagaagtttttgttagagtgtgtattactgtggaattcag
Jα45	tgtcagagaggtttatgtcaaggcttgcctcagggtggaattcag
Jα48	tggcacggtggattttgtaatggcttggcacactgtggaattcag
Jα49	tcttctgctggtttttgttgaggttattgtcacagtggaattcag
Jα50	cacctgccagtttttgtaaagggttgatttgctgtggaattcag
Jα52	tccacctggtgtttttgtaaaggcccgtgttacagtggaattcag
Jα53	cctaggtcctgtttttgtaaagcctccccaggctgtggaattcag
Jα56	ggacaactgagtttttgtagagtcccgtgtcattgtggaattcag
Jα57	aatctagatagtatttgtaaggcagtgtgtgggtgtggaattcag
Jα58	gatgccacgagtttttgcaaagcccttcagtgcagtggaattcag

3 Methods

3.1 Induction of aiCD4 T Cells in Mice

1. Mice (8 weeks old) are repeatedly immunized with 25 μg of SEB or PBS by means of i.p. injection every 5 days (*see* **Note 1**).
2. Nine days after the final immunization with SEB, SEB-reactive $V\beta8^+CD4^+$ T cells are isolated from the spleen using the $CD4^+$ T Cell Isolation Kit II.
3. This fraction is then treated with PE-conjugated anti-Vβ8 TCR antibody and separated by reaction with anti-PE microbeads to isolate the $V\beta8^+CD4^+$ T cell population.

3.2 Ligation-Mediated PCR (LM-PCR)

1. The BW-1 and BW-2 oligonucleotides (2 nmol each) are mixed in a volume of 100 μl of 250 mM Tris–HCl (pH 7.7). This mixture is heated to 90 °C for 5 min, and allowed to cool to room temperature to generate BW-1/2 linker mixture (*see* **Note 2**). This linker mixture is stored frozen, and can be thawed on ice before use.
2. Purified genomic DNA (1 μg) is ligated to 20 pmole of BW linker in a 50 μl of reaction buffer containing 66 mM Tris–HCl (pH 7.5), 5 mM $MgCl_2$, 5 mM DTT, 1 mM ATP, and 2.5 U of T4 DNA ligase (Roche Diagnostics, Basel, Switzerland) at 16 °C for 16 h.
3. The reaction mixture is diluted with an equal volume of a buffer containing 10 mM Tris (pH 8.3), 50 mM KCl, 0.5 % Nonidet P-40 (NP-40), and 0.5 % Tween 20.
4. This mixture is heated to 95 °C for 15 min to inactivate ligase, dissociate unligated linker strands, and to denature the DNA for subsequent PCR. Samples are stored at −20 °C until used for PCR.
5. A nested PCR strategy is used to identify the sites of linker ligation. Ligated DNA (100 ng) is used in a hot-start PCR assay containing 25 ng of the linker primer BW-1H and 25 ng of each locus-specific primer. These are added to a reaction mixture consisting of 10 mM Tris at pH 8.3, 50 mM KCl, 2 mM $MgCl_2$, and 0.5 units of Taq polymerase in a total volume of 25 μl. Samples are amplified for 12 cycles [1 min at 94 °C, 2 min at 60 °C], followed by a final 10 min extension step at 72 °C. The reaction product is then use to program an additional 27 cycles of PCR in an identical buffer containing a second, nested locus-specific primer and BW-1H as noted above (*see* **Note 4**).
6. The PCR products are analyzed by electrophoresis through 1.2 % agarose gels (*see* **Note 2**) and transferred to Hybond-N^+ membranes (GE Healthcare, Buckinghamshire, England).
7. Membranes are hybridized to alkaline phosphatase (ALP)-labeled probes (*see* **Note 3**) and visualized by Gene Images CDP-Star chemiluminescence reaction (GE Healthcare) (Fig. 1).

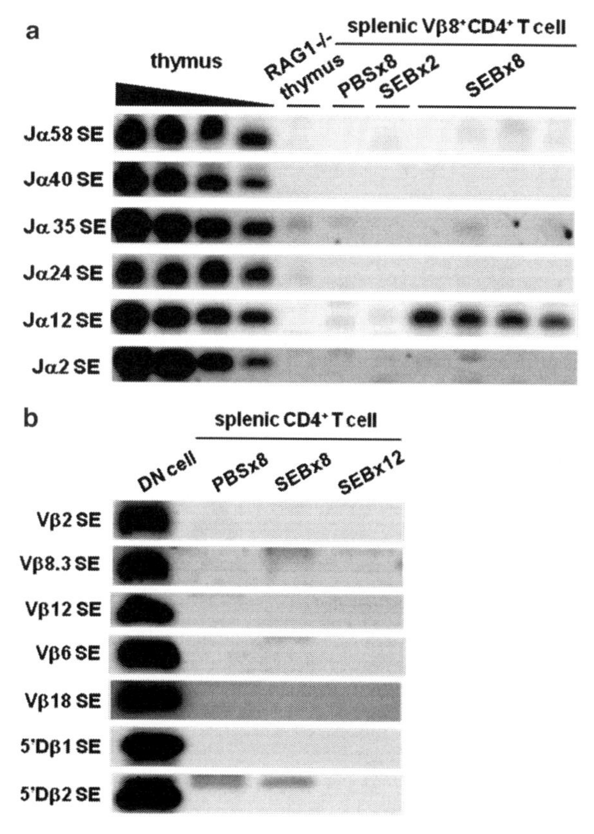

Fig. 1 The de novo TCRα revision upon repeated immunization with SEB. BALB/c mice were repeatedly immunized with 25 μg of SEB by means of i.p. injection every 5 days. SEB-reactive Vβ8+ CD4+ T cells or CD4+ T cells were isolated from spleen 9 days after final immunization. (**a**) TCRα revision was determined by using LM-PCR detection of the blunt signal ends (SEs), which are rearranged intermediates cleaved by RAG1/2 proteins, of TCRα joining (Jα) regions in SEB-reactive CD4+ T cells. (**b**) TCRβ revision determined by LM-PCR of the SEs of TCRβ variable (Vβ) and diversity (Dβ) regions in splenic CD4+ T cells. Double negative (DN) cell was isolated from thymus as a positive control of TCRβ revision

4 Notes

1. Basic immunization protocol can also be found in ref. 2.
2. To determine the exact location of the double-strand break, this amplicon is cloned, and individual colonies sequenced. For sequencing, purified PCR products isolated from agarose gels are cloned into the vector according to the manufacturer's instructions. In our laboratory, we used TOPO TA vector (Life technologies). This is then transformed into competent bacteria. After this, the plasmid DNA is sequenced using vector sequencing primers. The vector sequencing primers vary depending on the manufacturer and thus the reader is referred to respective manufacturers.

3. If chemiluminescence detection requires higher sensitivity, it may be increased by designing primers to generate probes of 300 bp or longer.
4. The BW-1H primer is identical to BW-1, the longer of the two strands comprising the linker, except for the addition of three nucleotides (CAG) at its 3′ end. These additional nucleotides are complementary to the first nucleotides of the RSS heptamer. By this technique, we can reduce the extent of nonspecific amplification, especially when we want to amplify low-frequency, linker-ligated targets.

References

1. Chervonsky AV (2010) Influence of microbial environment on autoimmunity. Nat Immunol 11:28–35
2. Tsumiyama K, Miyazaki Y, Shiozawa S (2009) Self-organized criticality theory of autoimmunity. PLoS One 4:e8382
3. Shiozawa S (2012) Pathogenesis of SLE and *ai*CD4 T cell: new insight on autoimmunity. Joint Bone Spine 79:428–430
4. Schlissel M, Constantinescu A, Morrow T, Baxter M, Peng A (1993) Double-strand signal sequence breaks in V(D)J recombination are blunt, 5′-phosphorylated, RAG-dependent, and cell cycle regulated. Genes Dev 7:2520–2532

Chapter 9

Basic Techniques for Studies of iNKT Cells and MAIT Cells

Asako Chiba and Sachiko Miyake

Abstract

Invariant natural killer T (iNKT) cells and mucosal-associated invariant T cells (MAIT cells) are T cell subsets belonging to innate-like lymphocytes. These innate-like lymphocytes express semi-invariant T cell receptors, but exert diverse functions and thus are involved in various types of immune responses. As iNKT cells and MAIT cells are abundant in human peripheral blood, these cells may hold important physiological roles, and thus it is desired to reveal their functions. Here, we first describe the cell preparation techniques commonly used in studies of innate-like lymphocytes, and then introduce methods for the detection and functional analysis of iNKT cells and MAIT cells.

Key words iNKT cells, MAIT cells, Innate-like lymphocytes, CD1d, MR1

1 Introduction

One of hallmark features of the adaptive immune cells is the diversity of their antigen receptors. MHC class I and II-restricted T cells express diverse T cell receptors (TCRs) and recognize peptide antigens. There are subsets of innate-like lymphocytes that also express TCRs. Unlike conventional MHC class I or II-restricted T cells, these innate-like lymphocytes including invariant natural killer T cells (iNKT cells) [1–5] and mucosal-associated invariant T cells (MAIT cells) [6–8] express semi-invariant TCRs and are restricted by nonclassical MHC class Ib molecules (Fig. 1). iNKT cells express a semi-invariant TCRα chain (Vα14-Jα18 in mice and Vα24-Jα18 in humans) with a limited set of TCRβ chains (Vβ8.2, Vβ7, Vβ2 in mice and Vβ11 in humans) and recognize lipid antigens including α-galactosylceramide (αGalCer) presented by the CD1d molecule. Recently, it has been revealed that there is another population of T cells expressing diverse TCRs restricted by the CD1d molecule. These CD1d-restricted T cells are referred to as type II NKT cells and also recognize lipid antigens, but not αGalCer. As little is known about type II NKT cells, we focus on iNKT cells in this

Shunichi Shiozawa (ed.), *Arthritis Research: Methods and Protocols*, Methods in Molecular Biology, vol. 1142, DOI 10.1007/978-1-4939-0404-4_9, © Springer Science+Business Media New York 2014

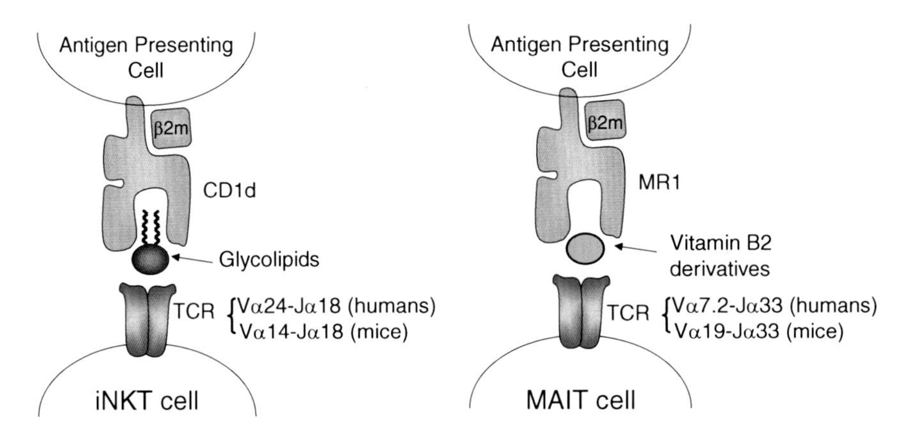

Fig. 1 Characteristics of iNKT cells and MAIT cells

chapter. MAIT cells also express a semi-invariant TCRα chain (Vα7.2-Jα33 in humans and Vα19-Jα33 in mice) and are restricted by the MHC-related molecule 1 (MR1).

Besides the invariant expression of antigen receptors, iNKT cells and MAIT cells possess distinctive characteristics of innate-like lymphocytes. These cells can be activated very rapidly not only by antigen receptor stimulation, but also by cytokines without undergoing clonal expansions in the periphery. Therefore, iNKT cells as well as MAIT cells seem to play important roles in the first-line immune responses. iNKT cells are known to be involved in various types of immune responses including infection, allergy, autoimmunity, and anticancer immunity. Several natural or synthetic CD1d ligands have been discovered to activate iNKT cells and modulate functions of these cells, and thus iNKT cell-targeted therapy is drawing attentions to clinical implications in the field of cancer, allergy, and autoimmune diseases. On the other hand, functions of MAIT cells have not been well characterized yet. MAIT cells were also reported to be involved in infection and autoimmunity [9–12]. iNKT cells account for 0.02 % in human peripheral blood, and recently it has been revealed that the frequency of MAIT cells is about 5 % among peripheral blood αβT cells in human [13]. Considering the fact that the approximate frequency of antigen specific T cell is one in one million αβT cells, these innate-like lymphocytes expressing semi-invariant TCRs are abundant and thus may hold important physiological roles. Therefore, further understating of these innate-like lymphocytes is desired.

As iNKT cells and MAIT cells also exist in tissues, it is required to investigate these cells not only in blood, but also in tissues including the spleen and liver. Therefore, in this chapter, first we describe common preparation methods of lymphocytes from blood and tissues. Next, we introduce methods to detect iNKT cells and MAIT cells by using FACS. We also explain methods for functional analysis of these innate-like lymphocytes.

2 Materials

2.1 General Materials for iNKT and MAIT Cell Studies

1. Phosphate buffered saline (PBS).
2. FACS buffer; PBS containing 0.1 % BSA or 2 % FCS.
3. Ficoll-Paque PLUS (GE Healthcare UK Ltd. Little Chalfont, Buckinghamshire, UK).
4. Percoll PLUS (GE Healthcare UK Ltd. Little Chalfont, Buckinghamshire, UK).
5. Ammonium Chloride–Potassium (ACK) Lysing Buffer.
6. Fc block (i.e., purified anti-mouse CD16/32 and purified anti-human Fc receptor blocking antibody).
7. Culture medium: RPMI 1640 or AIM V (Life Technologies, Carlsbad, CA, USA) supplemented with 10 % FCS, 2 mM L-GLUTAMINE, 50–100 U/ml of penicillin–streptomycin, and 50 mM β-mercaptoethanol.
8. Flow cytometer and/or Cell sorter.
9. Irradiator.

2.2 Materials for iNKT Cell Studies

2.2.1 Antibodies and CD1d Tetramers

1. Human or mouse CD1d tetramer (*see* **Note 1**).
   Both human and mouse unloaded CD1d tetramers conjugated with fluorochrome are commercially available. These tetramers need to be loaded with ligands for iNKT cell detection by FACS. Usually αGalCer or its derivatives are used as ligands. The antigen loading procedure should be performed following each manufacturer's instruction. CD1d tetramers loaded with PBS-57 or OCH are also available from NIH tetramer core facility. Store at 4 °C and protected from exposure to light.
2. Human or mouse Dimetric CD1d:Ig fusion protein (BD, Franklin Lakes, NJ, USA).
   Human and mouse Dimetric CD1d:Ig fusion proteins consists of extracellular domains of CD1d molecule fused with VH regions of mouse IgG1. A staining cocktail is made by mixing antigen-loaded Dimetric CD1d and anti-mouse IgG1 monoclonal antibody labeled with a fluorochrome. Store at 4 °C and protected from exposure to light.
3. Monoclonal antibody against human invariant NKT cell (Clone 6B11) (BD, Franklin Lakes, NJ, USA). Store at 4 °C and protected from exposure to light.
4. Monoclonal antibodies against human Vα24, Vβ11, CD3, CD4, and CD8 (BD, Franklin Lakes, NJ, USA). Store at 4 °C and protected from exposure to light.

2.2.2 CD1d Ligands (See Note 2)

Synthetic ligands: α-galactosylceramide (αGalCer) and its derivatives.

CD1d ligands are lipid and not soluble in water. The stock solution of ligand is usually prepared in DMSO, but the vehicle may change depending on the ligand. For the in vivo administration, stock solution of ligands is diluted with PBS containing Tween 80.

2.2.3 Materials for the Generation of Human iNKT Cell Lines

1. Human PBMC.
2. Recombinant cytokines: IL-2 and IL-7 (Peprotech, Rocky Hill, NJ, USA or R&D systems, Minneapolis, MN, USA).
3. PHA (Sigma-Aldrich, St. Louis, MO, USA).
4. AIM V medium (Life Technologies, Carlsbad, CA, USA) supplemented with 10 % FCS.
5. Monoclonal antibodies against human CD4, CD8, iNKT cell, Vα24, and Vβ11 (BD, Franklin Lakes, NJ, USA). Store at 4 °C and protected from exposure to light.

*2.2.4 Genetically Engineered Mouse Strains (See **Note 3**)*

1. Jα18 knockout mouse [14].
2. CD1d1 knockout mouse [15, 16].
3. Vα14-Jα281 transgenic (Tg) mouse [17, 18].

2.3 Materials for MAIT Cell Studies

MAIT cells are also distinguished by the expression of the invariant TCR. Lantz and colleagues generated monoclonal antibody against human Vα7.2, and revealed that Vα7.2^{positive}CD161^{high}αβT cells express invariant Vα7.2-Jα33 [13] (*see* **Note 4**). On the other hand, there are no specific antibodies against murine MAIT cell TCR. Although vitamin B metabolites were recently shown to be presented by the MR1 molecule and activate MAIT cells [19], but there is no MR1 tetramer available yet. Thus, genetic engineered mice are the available tools for murine MAIT cell studies.

2.3.1 Antibodies

Monoclonal antibodies against human CD3, Vα7.2, and CD161 (Biolegend, San Diego, CA).

*2.3.2 Genetically Engineered Mouse Strains (See **Note 5**)*

1. MR1 knockout mouse (C57BL6/J) [20].
2. Vα19TCR transgenic (tg) mouse (C57BL6/J) [13, 21].

3 Methods

3.1 Sample Preparation for iNKT and MAIT Cell Studies

3.1.1 PBMC Purification from Human Peripheral Blood

1. Dilute heparinized whole blood with the same amount of PBS.
2. Prepare 5 ml of Ficoll in a 15 ml-tube or 15 ml of Ficoll in a 50 ml-tube.
3. Overlay two volumes of diluted blood on Ficoll and centrifuge for 30 min at 680 ×*g*.
4. Transfer the interface lymphocyte layer to a tube and add 2–3 volumes of PBS.
5. Centrifuge at 470 ×*g* for 10 min and discard the supernatant.
6. After repeating **steps 4** and **5**, suspend the pellet with FACS buffer.

3.1.2 Isolation of Spleen Cells

1. Press spleen through a 70-μm cell strainer, and remove all cells by adding PBS onto the strainer.
2. Centrifuge at 470 × *g* for 5 min.
3. Discard the supernatant and add 1–2 ml of ACK buffer and incubate for 5–10 min.
4. Add three volumes of PBS and centrifuge at 470 × *g* for 5 min.
5. Remove the supernatant and repeat **step 4**.
6. Discard the supernatant and resuspend with FACS buffer or culture medium.

3.1.3 Isolation of Liver Mononuclear Cells

1. Perfuse a liver with PBS via portal vein, and then press liver fragments through a 70-μm cell strainer. Collect all cells on the strainer by adding PBS.
2. Centrifuge the liver cells in PBS for 5 min at 470 × *g* for 5 min.
3. Resuspend the pellet with 18 ml of 40 % Percoll.
4. Add 3 ml of 80 % Percoll and overlay 6 ml of the liver cells in 40 % Percoll.
5. Centrifuge at 1,600 × *g* for 30 min.
6. Collect the cells at the 40/80 % interface.
7. Wash once with FACS buffer and add 1 ml of ACK to lyse erythrocytes.
   After 1 min, wash the cells with FACS buffer twice.
8. Discard the supernatant and suspend the pellet with FACS buffer.

3.2 iNKT Cell Studies

3.2.1 Detection of iNKT Cells by FACS

As most iNKT cells express natural killer markers such as NK1.1, early studies on murine NKT cells were performed using NK1.1 and TCRβ. However, it is more appropriate to distinguish iNKT cells by the expression of the semi-invariant TCR. Human iNKT cells can be identified as CD1d tetramer^{+}TCRβ^{+}, CD1d dimer^{+}TCRβ^{+}, invariant (Vα24-JαQ) TCR^{+}, or Vα24^{+}Vβ11^{+}CD3^{+} cells. Mouse iNKT cells are also identified as CD1d tetramer^{+}TCRβ^{+} or CD1d dimer^{+}TCRβ^{+} cells.

1. Prepare single-cell suspensions in FACS staining buffer at a concentration of 0.5–5 × 10^6 cells/100 μl. The frequencies of mouse iNKT cells are about 20–30 % in the liver and 3 % in the spleen, and thus 1 × 10^6 cells are usually enough for mouse iNKT cell analysis. On the contrary, the frequency of iNKT cells in the human peripheral blood is very low and less than 0.1 % of T cells. Therefore, 3–5 × 10^6 PBMC per sample should be prepared.
2. Add Fc receptor blocking antibody and incubate more than 5 min prior to staining. Add FACS antibodies and/or CD1d tetramer to the cells and incubate on ice for 30 min.
3. Wash the cells with FACS buffer twice and keep cells in FACS buffer.
4. Analyze samples by FACS (*see* **Note 6**).

3.2.2 Activation of iNKT Cells by CD1d Ligands

iNKT cells become activated and produce cytokines such as IL-4, IL-13, IFNγ, and IL-17 upon stimulation.

1. In vivo activation of iNKT cells.
   An intravenous or intraperitoneal injection of αGalCer (0.1–0.5 mg/kg body weight) induces iNKT cells to produce cytokines, and this cytokine production occurs within a few hours after the αGalCer administration [22]. The elevation of serum IL-4 level is very rapid and observed as early as 2 h. IFNγ secreted by iNKT cells stimulate NK cells to produce IFNγ, thus the serum level of IFNγ reaches a plateau more than 12 h later. An administration of αGalCer-loaded dendritic cells induces iNKT cell activation in vivo [23]. An intravenous or intraperitoneal administration of OCH also activates iNKT cells in vivo, but in this case the elevation of serum IFNγ is limited [22].
2. In vitro activation of iNKT cells by CD1d ligands.
   For in vitro activation of iNKT cells, human PBMC, mouse splenocytes or liver MNCs are seeded onto 96-well plates and are cultured in medium containing CD1d ligands (e.g., 1–100 ng/ml of αGalCer). iNKT cell lines or NKT hybridoma are activated by ligands loaded onto plate-coated CD1d protein in the absence of antigen presenting cells [24, 25].

3.2.3 Generation of Human iNKT Cell Lines [26, 27]

Human PBMCs are stimulated with αGalCer (100 ng/ml) and cultured in media containing IL-2 (50 IU/ml) and IL-7 (10 ng/ml). Half of the medium is changed every 3–5 days with medium containing IL-2 (10UI/ml) and IL-7 (5 ng/ml). After 14–18 days, cells are labeled with anti-CD4, anti-CD8, anti-iNKT, and anti-Vβ11 mAbs and iNKT cells are sorted by Cell sorter. The sorted cells are cultured with allogeneic irradiated PBMC (100Gy) at a cell ratio of 1:3 and stimulated with 1.0 μg/ml PHA-P, IL-2 (50 IU/ml), and IL-7 (10 ng/ml) for 3 days and then maintained in basic medium supplemented with IL-2 (10 IU/ml) and IL-7 (5 ng/ml). Mouse iNKT cell lines can be also generated by using iNKT cells form from Vα14-Jα281 transgenic (Tg) mice [25] (*see* **Note 7**).

3.3 MAIT Cell Studies

3.3.1 Detection of Human MAIT Cells by FACS

1. Prepare single-cell suspensions in FACS staining buffer at a concentration of $0.5–5 \times 10^6$ cells/100 μl.
2. Add Fc receptor blocking antibody and incubate more than 5 min prior to staining.
3. Stain PBMC with monoclonal antibodies against CD3, TCRγδ, CD161, and Vα7.2 on ice for 30 min.
4. Wash the cells with FACS buffer twice and keep in FACS buffer.
5. Analyze samples by FACS (*see* **Note 8**).

3.3.2 Genetically Engineered Mice Strains

MAIT cells are known to be abundant in human peripheral blood and the gut lamina propria of humans and mice [13, 28]. However, the frequencies of MAIT cells are less in the mouse lymph nodes. MAIT cells expressing the invariant TCRα chain (Vα19-Jα33) are absent in MR1 knockout mice. Due to reduced frequency of MAIT cells in mice, sometimes it is required to use Vα19TCR transgenic (tg) mice to analyze function of MAIT cells.

4 Notes

1. CD1d tetramer: Unloaded CD1d tetramers can be obtained from MBL (Nagoya, Japan), Proimmune (Oxford, OX4 4GA, UK) or NIH tetramer core facility (Atlanta, GA, USA, http://tetramer.yerkes.emory.edu/). Human and mouse CD1d tetramers loaded with ligands (PBS-57 or OCH) are also available from NIH tetramer core facility.
2. CD1d ligands [18, 21, 29]:
   The first discovered CD1d ligand is αGalCer identified from a marine sponge and has been widely used in iNKT cell studies. Several derivatives of αGalCer have been synthesized and these derivatives hold different stimulatory capacity for iNKT cells. iNKT cells are autoreactive and thought to recognize endogenous ligands, but also respond to microbial ligands.

   Synthetic ligands: αCalCer, OCH, α-C-galactosylceramide, PBS-57

   Natural ligands: α-galacturonosylceramide (*Shongomonas* spp.), α-galacturonosylceramide (*Shongomonas* spp.), α-galactosyldiacylglycerol (*Borrelia burgdorferi*), phosphatidylinositol mannoside (*Mycobacterium bovis* BCG), house dust extracts.

   Endogenous ligands: isoglobotrihexosylceramide (iGb3), disialoganglioside, phosphatidylethanolamine, phosphatidylinositol, phosphatidylcholine.
3. Recently, it has been revealed that there are other T cells restricted by the CD1d molecule. These CD1d-restricted T cells express diverse TCRs and are called type II NKT cells (iNKT cells are also called type I NKT cells). Whereas Jα18 knockout mice lack iNKT cells, CD1d knockout mice lack both iNKT cells and type II NKT cells. There are two types of Vα14-Jα281 transgenic (Tg) mice available. One is on a wild-type C57BL6 mouse background, and another is on a TCRα-deficient background.
4. Martin et al. generated monoclonal antibody against human Vα7.2 TCR and showed that Vα7.2TCR positive αβT cells with high expression of CD161 are MAIT cells. They also

revealed that MAIT cells comprise up to 4 % among T cells in human peripheral blood. There is no antibody against mouse MAIT TCR yet.

5. MR1 deficient mice lack MAIT cells. It is not known whether there is another T cell population restricted by the MR1 molecule yet. Because the frequency of MR1-restricted Vα19i T cells in wild-type mice is low, Vα19i TCR-transgenic mice are useful to study function of mouse MAIT cells. Vα19i TCR-transgenic mice were generated by injecting C57BL/6J oocytes with a transgenic construct encoding Vα19-Jα33 TCR driven by *Tcra* promoter.
6. If the tetramer staining is not clear and nonspecific staining is observed, try to gate out B cells by co-staining with monoclonal antibody against a B cell surface marker such as CD19. The levels of CD3 or TCRβ expression on iNKT cells are low, thus antibodies conjugated with a bright fluorochrome are preferred. As antibodies against TCR may interfere with CD1d tetramer or dimer binding, it is important to test whether CD1d tetramer is feasible for co-staining with such antibodies.
7. Mouse iNKT cell lines: Although mouse iNKT cells are more difficult to grow in vitro, cell lines can be generated by using iNKT cells from Vα14-Jα281 transgenic (Tg) mice. The detailed method to generate these iNKT cell lines is described in reference [25]. These mouse iNKT cell lines possess characteristic features of iNKT cells, as they respond to various types of CD1d ligands and cytokine stimulation. In addition, these iNKT cells exerted functions in vivo when adoptively transferred into recipient mice [30].
8. MAIT cells are identified as $CD3^+TCR\gamma\delta^-V\alpha7.2^+CD161^{high}$ cells. The average frequency of MAIT cells in human PBMC is about 5 % among αβT cells. Most MAIT cells are $CD8^+$ or $CD4^-CD8^-$ double-negative cells. Human peripheral blood MAIT cells in the adults display an effector-memory phenotype ($CD45RA^-CD45RO^+CCR7^-CD62L^{lo}CD95^{hi}$) [31]. No specific antibody is available to distinguish murine MAIT cells yet.

Acknowledgement

This work was supported by a Grant-in-Aid for Scientific Research (B: 7210 to SM) from the Japan Society for the Promotion of Science.

References

1. Brigl M, Brenner MB (2004) CD1: antigen presentation and T cell function. Annu Rev Immunol 22:817–890
2. Godfrey DI, Kronenberg M (2004) Going both ways: immune regulation via CD1d-dependent NKT cells. J Clin Invest 114: 1379–1388
3. Bendelac A, Savage PB, Teyton L (2007) The biology of NKT cells. Annu Rev Immunol 25:297–336
4. Matsuda JL, Mallevaey T, Scott-Browne J, Gapin L (2008) CD1d-restricted iNKT cells, the 'Swiss-Army knife' of the immune system. Curr Opin Immunol 20:358–368
5. Brennan PJ, Brigl M, Brenner MB (2013) Invariant natural killer T cells: an innate activation scheme linked to diverse effector functions. Nat Rev Immunol 13:101–117
6. Treiner E, Lantz O (2006) CD1d- and MR1-restricted invariant T cells: of mice and men. Curr Opin Immunol 18:519–526
7. Gapin L (2009) Where do MAIT cells fit in the family of unconventional T cells? PLoS Biol 31:e70
8. Le Bourhis L, Guerri L, Dusseaux M, Martin E, Soudais C, Lantz O (2011) Mucosal-associated invariant T cells: unconventional development and function. Trends Immunol 32:212–218
9. Croxford JL, Miyake S, Huang YY, Shimamura M, Yamamura T (2006) Invariant Vα19i T cells regulate autoimmune inflammation. Nat Immunol 7:987–994
10. Le Bourhis L, Martin E, Péguillet I, Guihot A, Froux N, Coré M, Lévy E, Dusseaux M, Meyssonnier V, Premel V, Ngo C, Riteau B, Duban L, Robert D, Huang S, Rottman M, Soudais C, Lantz O (2010) Antimicrobial activity of mucosal-associated invariant T cells. Nat Immunol 11:701–718
11. Gold MC, Cerri S, Smyk-Pearson S, Cansler ME, Vogt TM, Delepine J, Winata E, Swarbrick GM, Chua WJ, Yu YY, Lantz O, Cook MS, Null MD, Jacoby DB, Harriff MJ, Lewinsohn DA, Hansen TH, Lewinsohn DM (2010) Human mucosal associated invariant T cells detect bacterially infected cells. PLoS Biol 29:e1000407
12. Chiba A, Tajima R, Tomi C, Miyazaki Y, Yamamura T, Miyake S (2012) Mucosal-associated invariant T cells promote inflammation and exacerbate disease in murine models of arthritis. Arthritis Rheum 64:153–161
13. Martin E, Treiner E, Duban L, Guerri L, Laude H, Toly C et al (2009) Stepwise development of MAIT cells in mouse and human. PLoS Biol 7:e54
14. Cui J, Shin T, Kawano T, Sato H, Kondo E, Toura I, Kaneko Y, Koseki H, Kanno M, Taniguchi M (1997) Requirement for Valpha14 NKT cells in IL-12-mediated rejection of tumors. Science 278:1623–1626
15. Chen YH, Chiu NM, Mandal M, Wang N, Wang CR (1997) Impaired NK1+ T cell development and early IL-4 production in CD1-deficient mice. Immunity 6:459–467
16. Exley MA, Bigley NJ, Cheng O, Shaulov A, Tahir SM, Carter QL, Garcia J, Wang C, Patten K, Stills HF, Alt FW, Snapper SB, Balk SP (2003) Innate immune response to encephalomyocarditis virus infection mediated by CD1d. Immunology 110:519–526
17. Taniguchi M, Koseki H, Tokuhisa T, Masuda K, Sato H, Kondo E, Kawano T, Cui J, Perkes A, Koyasu S, Makino Y (1996) Essential requirement of an invariant V alpha 14T cell antigen receptor expression in the development of natural killer T cells. Proc Natl Acad Sci U S A 93:11025–11028
18. Kawano T, Cui J, Koezuka Y, Toura I, Kaneko Y, Motoki K, Ueno H, Nakagawa R, Sato H, Kondo E, Koseki H, Taniguchi M (1997) CD1d-restricted and TCR-mediated activation of valpha14 NKT cells by glycosylceramides. Science 278:1626–1629
19. Kjer-Nielsen L, Patel O, Corbett AJ, Le Nours J, Meehan B, Liu L, Bhati M, Chen Z, Kostenko L, Reantragoon R, Williamson NA, Purcell AW, Dudek NL, McConville MJ, O'Hair RA, Khairallah GN, Godfrey DI, Fairlie DP, Rossjohn J, McCluskey J (2012) MR1 presents microbial vitamin B metabolites to MAIT cells. Nature 491:717–723
20. Treiner E, Duban L, Bahram S, Radosavljevic M, Wanner V, Tilloy F, Affaticati P, Gilfillan S, Lantz O (2003) Selection of evolutionarily conserved mucosal-associated invariant T cells by MR1. Nature 422:164–169
21. Okamoto N, Kanie O, Huang YY, Fujii R, Watanabe H, Shimamura M (2005) Synthetic α-mannosyl ceramide as a potent stimulant for an NKT cell repertoire bearing the invariant Vα19-Jα26 TCR alpha chain. Chem Biol 12:677–683
22. Miyamoto K, Miyake S, Yamamura T (2001) A synthetic glycolipid prevents autoimmune encephalomyelitis by inducing TH2 bias of natural killer T cells. Nature 413:531–534
23. Fujii S, Shimizu K, Kronenberg M, Steinman RM (2002) Prolonged IFN-γ-producing NKT response induced with α-galactosylceramide-loaded DCs. Nat Immunol 3:867–874
24. Gumperz JE, Roy C, Makowska A, Lum D, Sugita M, Podrebarac T, Koezuka Y, Porcelli

SA, Cardell S, Brenner MB, Behar SM (2000) Murine CD1d-restricted T cell recognition of cellular lipids. Immunity 12:211–221

25. Chiba A, Cohen N, Brigl M, Brennan PJ, Besra GS, Brenner MB (2009) Rapid and reliable generation of invariant natural killer T-cell lines in vitro. Immunology 128:324–333
26. Araki M, Kondo T, Gumperz JE, Brenner MB, Miyake S, Yamamura T (2003) Th2 bias of CD4+ NKT cells derived from multiple sclerosis in remission. Int Immunol 15:279–288
27. Sakuishi K, Oki S, Araki M, Porcelli SA, Miyake S, Yamamura T (2007) Invariant NKT cells biased for IL-5 production act as crucial regulators of inflammation. J Immunol 179:3452–3462
28. Tilloy F, Treiner E, Park SH, Garcia C, Lemonnier F, de la Salle H, Bendelac A, Bonneville M, Lantz O (1999) An invariant T cell receptor alpha chain defines a novel TAP-independent major histocompatibility complex class Ib-restricted alpha/beta T cell subpopulation in mammals. J Exp Med 189:1907–1921
29. Godfrey DI, Rossjohn J (2011) New ways to turn on NKT cells. J Exp Med 208: 1121–1125
30. Sada-Ovalle I, Chiba A, Gonzales A, Brenner MB, Behar SM (2008) Innate invariant NKT cells recognize Mycobacterium tuberculosis-infected macrophages, produce interferon-gamma, and kill intracellular bacteria. PLoS Pathog 4:e1000239
31. Dusseaux M, Martin E, Serriari N, Péguillet I, Premel V, Louis D, Milder M, Le Bourhis L, Soudais C, Treiner E, Lantz O (2011) Human MAIT cells are xenobiotic-resistant, tissue-targeted, CD161hi IL-17-secreting T cells. Blood 117:1250–1259

Chapter 10

Induction of De Novo Autoimmune Disease in Normal Mice upon Repeated Immunization with Antigen

Ken Tsumiyama and Shunichi Shiozawa

Abstract

There are many issues with animal models that represent human autoimmune disease or protocols to induce systemic autoimmunity, especially protocols to induce disease in normal mice not having a genetic disposition to autoimmunity. We describe here a novel and completely reproducible experimental technique that can induce systemic autoimmunity or systemic lupus erythematosus (SLE) in mice otherwise not prone to spontaneous autoimmune disease. This protocol involves the repeated immunization of mice with the same antigen. This rather simple technique enables us to perform exact and quantitative in vivo animal experiments with great accuracy.

Key words Systemic autoimmunity, Systemic lupus erythematosus (SLE), Animal model, Repeated immunization with antigen, Autoantibodies, Immune tissue injury

1 Introduction

Attempts to experimentally induce autoimmunity or systemic lupus erythematosus (SLE) in normal animals, i.e., without an autoimmune genetic background, have been unsuccessful [1]. We therefore took an approach that departed from the traditional concept of "autoimmunity," and tried to view this pathogenesis from a different angle, i.e., the integrity of immune system. The method we have chosen is to maximally stimulate the immune system with antigen to the levels far beyond its normal capacity, akin to testing the capability of automobile in an extreme condition such as an F1 car race. We found that when we repeatedly immunized mice with the same antigen we inevitably induced systemic autoimmunity. We theorize that autoimmunity occurs when the host's immune system is overstimulated by repeated exposure to antigen to levels that surpass some inherent homeostatic level, i.e., the immune system's self-organized criticality [2, 3]. In this report, we describe this experimental technique in detail. The experimental technique shown here is simple but reproducible, thereby enabling one to perform quantitative animal in vivo experiments with great accuracy.

Shunichi Shiozawa (ed.), *Arthritis Research: Methods and Protocols*, Methods in Molecular Biology, vol. 1142, DOI 10.1007/978-1-4939-0404-4_10, © Springer Science+Business Media New York 2014

2 Materials

Antigen is repeatedly injected into mice without adjuvant, as it has been suggested that adjuvant stimulation itself may induce autoimmunity [4, 5].

2.1 Repeated Immunization with Antigen

1. Eight week-old female BALB/c mice (Japan SLC, Hamamatsu, Japan).
2. Staphylococcal enterotoxin B (SEB; Toxin technology, Sarasota, FL, USA) (*see* **Note 1**).
3. Keyhole limpet hemocyanin (KLH; Sigma, St. Louis, MO, USA) (*see* **Note 1**).
4. Ovalbumin (OVA; grade V; Sigma) (*see* **Note 1**).
5. Distilled phosphate buffered saline (PBS; Nissui, Tokyo, Japan).
6. 1 ml syringe and 26-gauge needle (TERUMO COOPERATION, Tokyo, Japan).
7. IgM- or IgG-rheumatoid factor (RF) mouse ELISA kit (Sibayagi Co., Gunma, Japan).
8. Sm antigen (ImmunoVision, Springdale, AR, USA).
9. Calf thymus-derived DNA (Worthington Biochemical Co., Lakewood, NJ, USA).
10. S1 nuclease (Promega, Madison, WI, USA).
11. Sera of prototypic autoimmune MRL/lpr mice.

3 Methods

Animal studies are approved by the institutional animal care and use committee, and carried out in accord with appropriate institutional regulations. In contrast to previous animal studies, the experimental protocol shown here is completely repeatable, as long as the appropriate antigens and mice are used.

3.1 Induction of SLE by Repeated Immunization with Antigen

1. Dissolve SEB with PBS to a concentration of 100 μg/ml.
2. Dissolve KLH with PBS to a concentration of 400 μg/ml.
3. Dissolve OVA with PBS to a concentration of 2 mg/ml.
4. Immunize BALB/c mice with 250 μl of 100 μg/ml SEB (25 μg of SEB per mouse), 250 μl of 400 μg/ml KLH (100 μg of KLH per mouse) or 250 μl of 2 mg/ml OVA (500 μg of OVA per mouse) by means of intraperitoneal (i.p.) injection using a 1 ml syringe and 26-gauge needle.
5. Immunize with antigen repeatedly as in **step 4** for a total of 8–12 times every 5 days (*see* **Note 2**).
6. Collect sera 2 days after every second immunization to detect autoantibodies using enzyme-linked immunosorbent assay

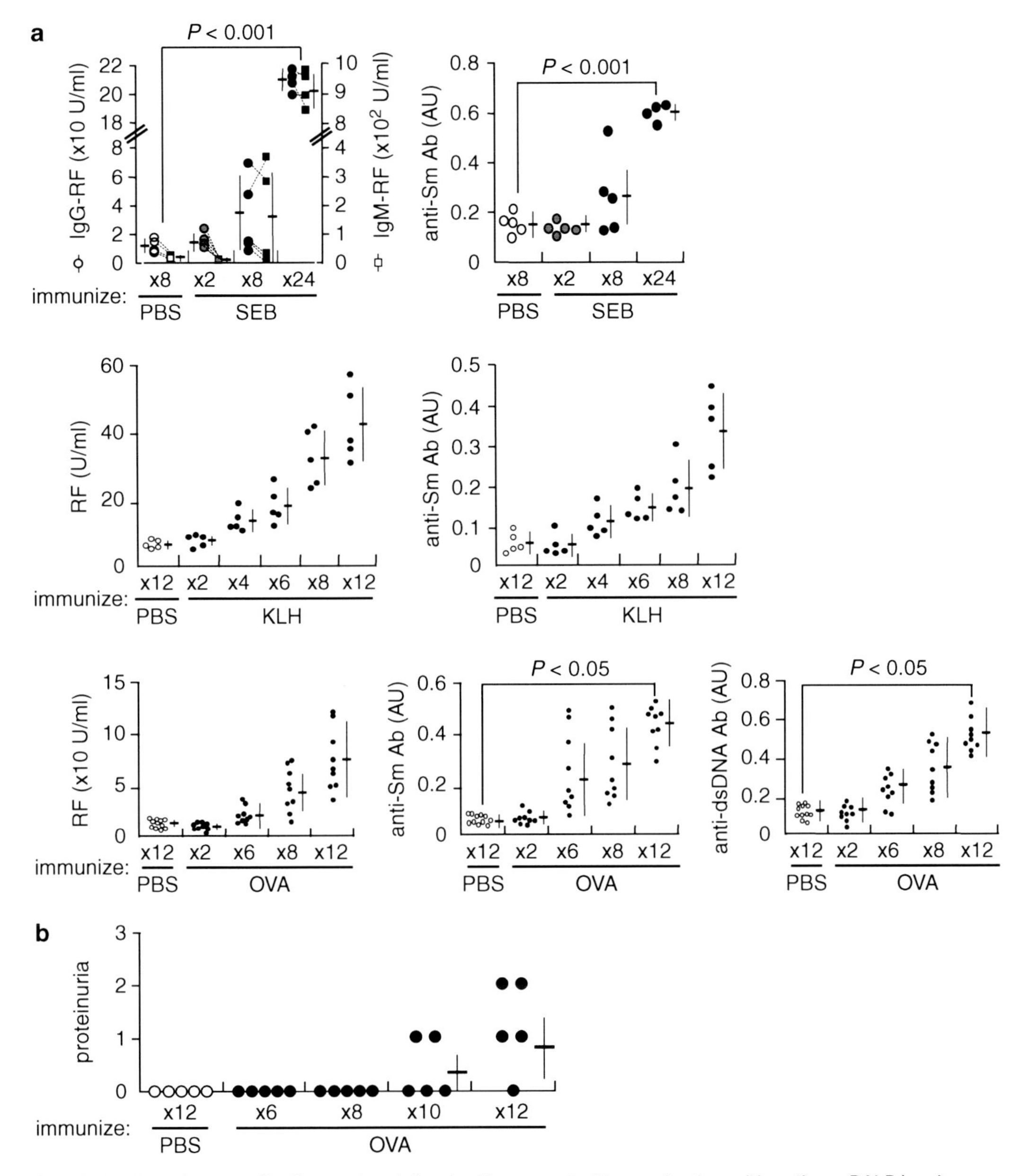

Fig. 1 Induction of autoantibodies and proteinuria after repeated immunization with antigen. BALB/c mice were repeatedly injected i.p. with 25 μg of SEB, 100 μg of KLH or 500 μg of OVA every 5 days. (**a**) Serum RF, anti-Sm, and anti-dsDNA antibodies were quantified by ELISA 2 days after each immunization. The arbitrary unit (AU) of 1.0 is defined as the titer obtained from the sera of MRL/lpr mice. Data from each mouse are connected by *dotted lines*. (**b**) Proteinuria was assessed 2 days after each immunization with OVA and graded with a score of 0 (<30 mg/dl); 1 (30–100 mg/dl); 2 (100–300 mg/dl); or 3 (300–1,000 mg/dl)

(ELISA). RF is detected using IgM- or IgG-RF mouse ELISA kit according to manufacturer's instructions. Anti-Sm and anti-double-stranded DNA (dsDNA) antibodies are also detected using Sm antigen and calf thymus-derived DNA digested S1 nuclease (*see* **Note 3**). We used sera of MRL/lpr mice as a positive control (*see* **Notes 4** and **5**) (Fig. 1a).

7. Two days after every second immunization, collect urine to detect proteinuria (*see* **Notes 6** and **8**) (Fig. 1b).
8. Nine days after final immunization, collect tissues and proceed to pathological studies (*see* **Notes** 7 and **8**) (Fig. 2).

4 Notes

1. Repeated immunization with antigen induces autoantibodies and immune tissue injury, regardless of the immunizing antigen lot. These antigens are injected i.p. into mice without any adjuvant [4, 5].
2. The immunization interval is not related for to induction of autoantibodies. Mice that were immunized with SEB every 15 days for a total eight times showed similar induction of RF as seen in mice immunized with SEB every 5 days for a total of eight times.
3. To prepare dsDNA, 5 mg of calf thymus-derived DNA (2.5 ml of 2 mg/ml DNA) is incubated with 500 units of S1 nuclease at 37 °C for 30 min. We coat 96-well plates (Maxisoup; Nunc, Roskilde, Denmark) with 100 μl/well of 10 units/ml Sm antigen diluted with PBS or 500 μg/ml dsDNA dissolved with 1× SSC buffer (0.15 M of NaCl and 15 mM of sodium citrate) at 4 °C for overnight. After blocking with 2 % bovine serum albumin (BSA; Sigma) in PBS at room temperature for 1 h, 100 μl of sera diluted 50-fold with PBS are added to the wells, and the plate is incubated at 4 °C for overnight or at room temperature for 2 h to detect anti-Sm or anti-dsDNA antibody, respectively.
4. Any antigen can induce autoantibodies by repeated immunization, and autoantibodies usually begin to arise after 8× immunization (Fig. 1a).
5. Induction of autoantibodies does require correct antigen presentation. SEB is efficiently presented through the I-E molecule of the H-2 antigen. When SEB is repeatedly immunized to BALB/c and B10.D2 mice, both of which are I-E molecule-positive, RF is consistently induced. However, similar immunization of C57BL/6 mice, which do not have I-E molecule, fails to induce RF. Thus, as long as the immunizing antigen is correctly presented on a major histocompatibility complex (MHC) to CD4 T cells, autoantibodies can be generated upon repeated immunization in all otherwise healthy mice [2, 3].
6. In our laboratory, proteinuria is detected using test strips (Albustix; Siemens Healthcare Diagnostics Inc., Tarrytown, NY, USA) according to the manufacturer's instructions [2]. Proteinuria is usually detectable in about 40 and 80 % of the mice after 10× and 12× immunization with OVA, respectively (Fig. 1b).

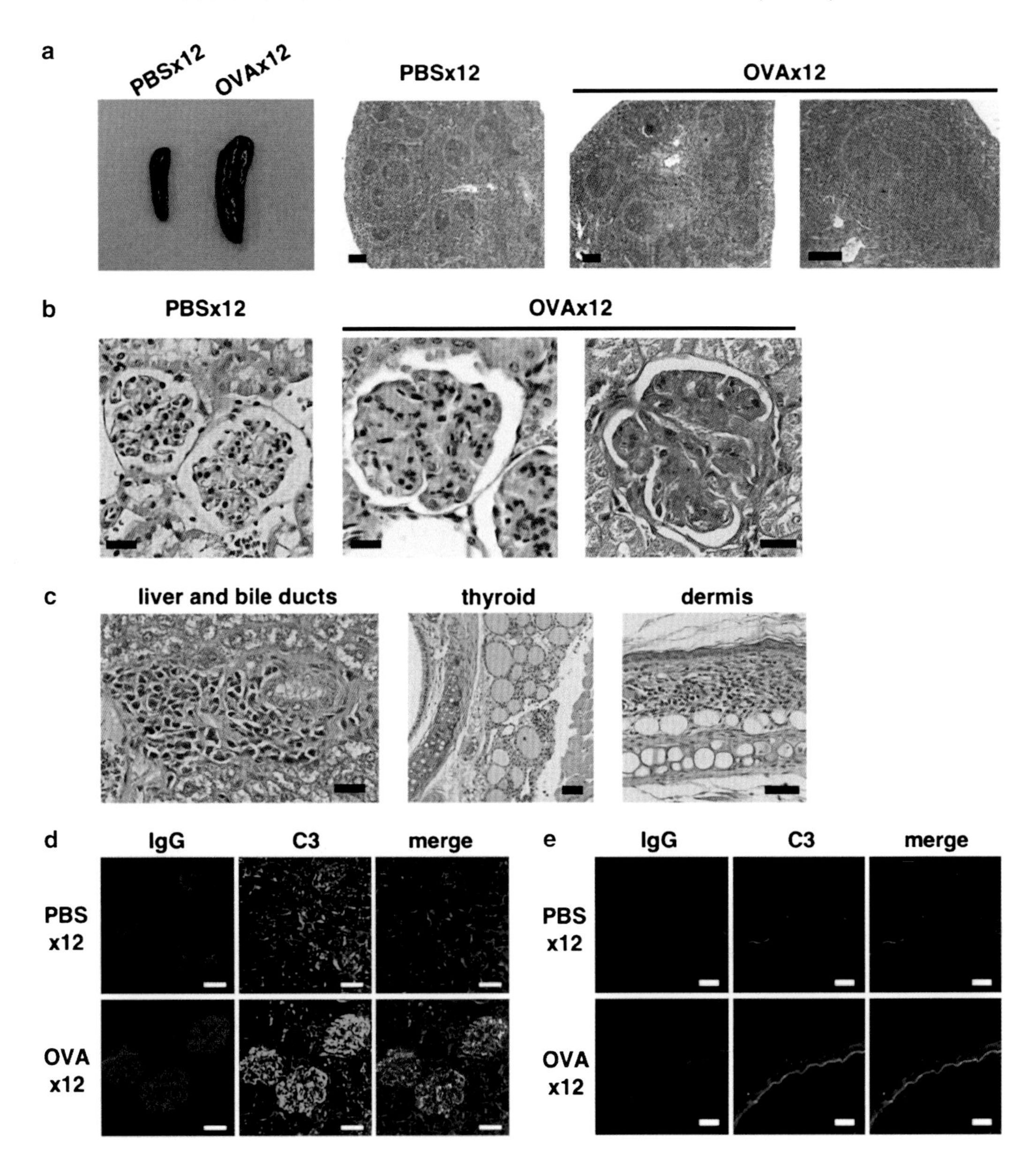

Fig. 2 Histopathology of mice repeatedly immunized with antigen. BALB/c mice were repeatedly injected i.p. with 500 μg of OVA every 5 days. Pathological changes were assessed 9 days after immunization 12× with OVA. (**a**) Splenomegaly after immunization 12× with OVA. Enlarged lymphoid follicles with marked germinal centers were seen in mice (H&E staining, bar = 200 μm; original magnification ×20). (**b**) Representative renal histopathology in the mice immunized 12× with OVA. Glomerular expansion with cellular infiltration including eosinophils was seen (*middle*, H&E staining, bar = 20 μm; original magnification ×400). A wire loop-like massive membranous glomerulonephritis in the kidney (*right*, PAS staining, bar = 20 μm; original magnification ×400). (**c**) Plasma cell infiltrates around bile ducts (*left*) (H&E staining, bar = 20 μm; original magnification ×400), focal infiltrates of mononuclear cells into the thyroid (*middle*) (H&E staining, bar = 50 μm; original magnification ×100), and diffuse infiltration of inflammatory cells into auricular subcutaneous tissue (*right*, H&E staining, bar = 50 μm; original magnification ×200). (**d**) Immunofluorescent staining for deposited IC, IgG, and C3 (bar = 50 μm; original magnification ×200). (**e**) Lupus band test stained with anti-IgG and anti-C3 antibodies (bar = 20 μm; original magnification ×400)

7. We examine pathological changes using several kinds of tissue staining techniques, including hematoxylin and eosin (H&E), periodic acid-Schiff (PAS), or immunofluorescent staining. We also perform the "lupus band test," diagnostic for SLE, by detecting the deposition of immune complex (IC), IgG and C3 in the skin specimens of mice using antibodies such as goat anti-C3 antibody (Bethyl laboratories, Inc., Montgomery, TX, USA), Alexa Fluor 488-conjugated rabbit anti-goat IgG, and Alexa Fluor 594-conjugated rabbit anti-mouse IgG (Life Technologies Cooperation, Carlsbad, CA, USA). Specimens are examined by fluorescence microscopy [2] (Fig. 2).
8. While repeated immunization with OVA induces immune tissue injury in mice, proteinuria and pathological changes are not induced upon repeated immunization with SEB or KLH. Only antigens that can be cross-presented to CD8 T cell to induce final differentiation into cytotoxic T lymphocyte (CTL) can induce immune tissue injury. The causative antigen can be individually different; however, an antigen that can induce autoantibodies and tissue injury must be correctly presented and cross-presented to CD4 and CD8 T cells in the context of MHC [2, 3].

References

1. Perry D, Sang A, Yin Y, Zheng YY, Morel L (2011) Murine models of systemic lupus erythematosus. J Biomed Biotechnol 2011:271694
2. Tsumiyama K, Miyazaki Y, Shiozawa S (2009) Self-organized criticality theory of autoimmunity. PLoS One 4:e8382
3. Shiozawa S (2012) Pathogenesis of SLE and *ai*CD4 T cell: new insight on autoimmunity. Joint Bone Spine 79:428–430
4. Kuroda Y, Nacionales DC, Akaogi J, Reeves WH, Satoh M (2004) Autoimmunity induced by adjuvant hydrocarbon oil components of vaccine. Biomed Pharmacother 58:325–337
5. Rose NR (2008) The adjuvant effect in infection and autoimmunity. Clin Rev Allergy Immunol 34:279–282

Chapter 11

Mouse Model of Experimental Dermal Fibrosis: The Bleomycin-Induced Dermal Fibrosis

Jérôme Avouac

Abstract

Relevant animal models are essential tools to investigate in depth the pathogenesis of autoimmune disease. Systemic sclerosis (SSc) is an autoimmune connective tissue disorder that affects particularly the skin. SSc is characterized by vasculopathy, immune disturbances, and fibrosis. Expression of each of the three pathologic features varies among SSc patients leading to disease heterogeneity and variable organ manifestations. Several animal models of SSc are available; however, some models display inflammation followed by fibrosis, whether some others primarily mimic autonomous fibroblast activation. Here, we describe the mouse model of bleomycin-induced dermal fibrosis, which mimics early and inflammatory stages of SSc, and is widely used in SSc research.

Key words Systemic sclerosis, Dermal fibrosis, Animal model, Inflammation, Bleomycin

1 Introduction

Bleomycin-induced dermal fibrosis is a model of the early and inflammatory stages of systemic sclerosis (SSc), a connective tissue disease of unknown etiology that affects particularly the skin. Early stages of SSc are characterized by vascular changes and inflammatory infiltrates in the lesional skin [1]. Later stages of SSc are characterized by an excessive accumulation of extracellular matrix components, including collagen, leading to increased skin thickness. Bleomycin, originally isolated from *Streptomyces verticillus*, is a cancer chemotherapy that has established itself as a potent initiator of tissue injury and fibrosis in mice and other animal species. It has become associated with models of dermal and lung fibrosis that have become a cornerstone of experimental biology that is highly relevant to studies of SSc.

Starting from Yamamoto and coworkers, who established a murine model for skin fibrosis by daily subcutaneous injections of bleomycin over a 4-week period [2], several protocols have been proposed to induce bleomycin-induced dermal fibrosis [3–8].

Shunichi Shiozawa (ed.), *Arthritis Research: Methods and Protocols*, Methods in Molecular Biology, vol. 1142, DOI 10.1007/978-1-4939-0404-4_11,

Repetitive bleomycin injections induce skin fibrosis localized to the area surrounding the injecting site. The presence of anti-topoisomerase-I, anti-U1-RNP, and anti-histone antibodies may also reflect systemic disease involvement in this model [9, 10].

Histopathological examination revealed definite dermal fibrosis characterized by thickened collagen bundles and the deposition of homogeneous materials in the thickened dermis with cellular infiltrates, which mimics the histological features of human SSc. Bleomycin treatment induces the production of reactive oxygen species, causes damage to endothelial cells and other cell types, and leads to the expression of adhesion molecules [11]. This attracts leukocytes, including T cells, monocytes/macrophages, and mast cells, which are supposed to play an important role in the induction of fibrosis. These inflammatory cells infiltrate into lesional skin and activate resident fibroblasts [2]. Activated fibroblasts then produce and release large amounts of extracellular matrix, which results in skin fibrosis at the site of bleomycin injection. In addition to tissue infiltration by leukocytes, the model of bleomycin induced skin fibrosis exhibits an early and sustained activation of profibrotic TGF-β signaling [12–14]. Apart from TGF-β, a broad variety of other inflammatory and profibrotic mediators, including monocyte chemoattractant protein-1, PDGF, IL-4, IL-6, and IL-13, is overexpressed in skin challenged with bleomycin [11].

2 Materials

2.1 Induction of Bleomycin-Induced Dermal Fibrosis

1. Characteristics of the mice used in this model: 6-week-old mice (*see* **Notes 1** and **2**).
2. Bleomycin is dissolved in 0.9 % sodium chloride (NaCl) at a concentration of 0.5 mg/ml, and sterilized by filtration (0.2 μm).
3. 0.9 % NaCl.
4. 1 ml syringes.
5. 27-gauge needles.

2.2 Mouse Sacrifice, Blood Collection, and Skin Removal

1. Mouse sacrifice: Ethanol, 70 % in a squeeze bottle, absorbent paper.
2. Blood collection and skin removal:
   (a) 3 mm skin biopsy punch Stiefel®.
   (b) Tissue forceps.
   (c) 13 cm dressing scissors.
   (d) 4 % paraformaldehyde (PFA).
   (e) RNAlater (Ambion, Saint Aubin, France).
   (f) Ethanol, 70 %.

(g) Liquid nitrogen.

(h) 1 ml syringes and 22-gauge needles.

(i) Tissue cassettes, cryotubes, and eppendorfs.

3 Methods

3.1 Injection Protocol

1. Two operators are needed for the injection protocol: one operator holds the mouse while the second shaves the mouse, draws the square, and performs the injection.
2. Shave the upper back of the mice with a mechanical or electric shaver. Avoid shaving the bottom back, which is more subjected to mechanical stress.
3. Draw a square of 1 cm^2 in the shaved upper back area with a permanent marker.
4. Perform a subcutaneous injection of 100 μl of bleomycin dissolved in 0.9 % NaCl at a concentration of 0.5 mg/ml. The first four subcutaneous injections should be performed each time in a different corner of the square, and the fifth in the middle of the square (Fig. 1).
5. Subcutaneous injections of 100 μl 0.9 % NaCl were used as controls.
6. Local injections of bleomycin are performed daily for 3 weeks.

3.2 Mouse Sacrifice and Skin Removal

The sacrifice of the mice should be performed the next day after the final bleomycin injection.

1. Mouse sacrifice by cervical dislocation: Place the mouse on top of a cage, so that it grips the bars with its front paws. Break its

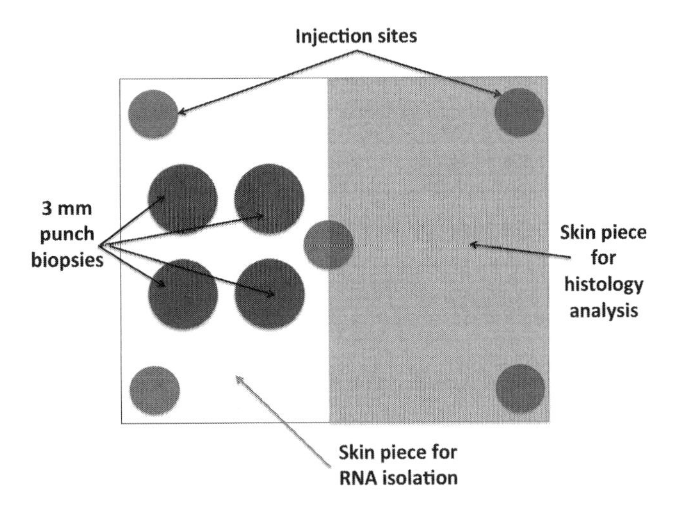

Fig. 1 Representation of the lesional skin area injected with bleomycin

neck humanely by applying firm pressure at the base of the skull while at the same time pulling backward on the tail. Alternatively, apply pressure to the base of the skull with a spatula, pencil, or cage cardholder. The CO_2 inhalation method also may be used if necessary. Lay the animal on its back on absorbent paper and soak it thoroughly in 70 % ethanol from a squeeze bottle. This important step reduces the risk of contaminating the dissection with mouse hair.

2. Blood collection, serum collection, and skin removal
   Blood collection by cardiac puncture: Hold the mouse by the scruff of skin above the shoulders so that its head is up and its rear legs are down. Use a 1 ml syringe and a 22-gauge needle. Insert needle 5 mm from the center of the thorax towards the animal's chin, 5–10 mm deep, holding the syringe 25–30° away from the chest. If blood does not appear immediately, withdraw 0.5 cc of air to create a vacuum in the syringe. Withdraw the needle without removing it from under the skin and try a slightly different angle or direction. When blood appears in the syringe, hold it still and gently pull back on the plunger to obtain the maximum amount of blood available. Pulling back on the plunger too much will cause the heart to collapse. If blood stops flowing, rotate the needle or pull it out slightly.

   Serum collection: untreated tubes should be used, as serum is collected in the absence of anticoagulant. Keep blood at room temperature for 1 h in the untreated collection tubes, then centrifuge the tubes for 15 min at 3,000 rpm ($1,500 \times g$), 4 °C.

   Skin removal: Carefully cut the 1 cm^2 square of skin with the iris scissors. The skin piece is then cut into two (Fig. 1). The first piece is fixed in 4 % PFA for 24 h and is placed in a tissue cassette in 70 % ethanol. This fragment will be further embedded in paraffin for histology and immunohistochemistry. Four skin biopsies are performed in the second piece using 3 mm biopsy punch. Skin biopsies are then snap-frozen in liquid nitrogen and stored immediately at −80 °C. The remaining skin of the second fragment is stored in RNAlater at −80 °C (*see* **Note 3**).

4 Notes

1. Bleomycin-induced dermal fibrosis has previously been induced in various mice strains, although there was some variation among strains in the intensity of the symptoms and the period required to induce dermal sclerosis; C3H/He, DBA/2, B10.D2, and B10.A mice, for example, show high susceptibility to bleomycin-induced fibrosis [15] (Table 1).
2. Males and females can be used with this model, although males show higher susceptibility to bleomycin-induced fibrosis than females.

Table 1
Bleomycin-induced dermal fibrosis

Features	Advantages	Limitations
Absence of vasculopathy Inflammation: ++ Autoimmunity: ++ Fibrosis: +++	Well described Applicable to many different mouse strains Easy handling	Artificial No major systemic disease involvement Overestimates the effects of anti-inflammatory drugs

3. Three different validated outcome measures are usually used to assess fibrosis in this model. The measure of dermal thickness is performed on 5 μm thick paraffin-embedded skin sections, after staining with hematoxylin and eosin (Fig. 2). The dermal thickness is analyzed at 100-fold magnification by measuring the distance between the epidermal–dermal junction and the dermal–subcutaneous fat junction at four sites from lesional skin of each mouse. The collagen content in lesional skin samples can be explored by the Masson's Trichrome staining (Sigma Aldrich, Saint-Quentin Fallavier, France) (Fig. 3) or can be estimated by determining the total content of hydroxyproline in lesional skin [16]. The hydroxyproline assay is performed on the punch biopsies (Ø 3 mm). The numbers of myofibroblasts are quantified by immunohistochemistry staining for α-smooth muscle actin (α-SMA) in paraffin-embedded sections [3–5] (Fig. 4).
4. The model of bleomycin-induced skin fibrosis is relatively easy to perform and applicable to most mouse strains. It is useful to evaluate anti-inflammatory and anti-fibrotic therapies in preclinical studies of SSc. This model can be used for the prevention or the treatment of fibrosis. Regarding the preventive approach, treatment with a candidate molecule starts prior or at the onset of bleomycin injections. To assess the treatment of established fibrosis, injections of the pharmacologic treatment start several weeks (in general 3 weeks) after the onset of bleomycin injections. In this modified protocol, novel therapeutics may not only stop progression but also induce regression of established fibrosis.
5. Although bleomycin-induced skin fibrosis is widely used as a model of early inflammatory stages of SSc, there are some important limitations. At least with usual doses of bleomycin, fibrosis is limited to the site of injection, whereas SSc is a systemic disease involving the skin and internal organs. Typical generalized microangiopathy that precedes fibrosis in human SSc is usually absent in the bleomycin model. Nevertheless, thickness of vascular wall in the deep dermis can be observed.

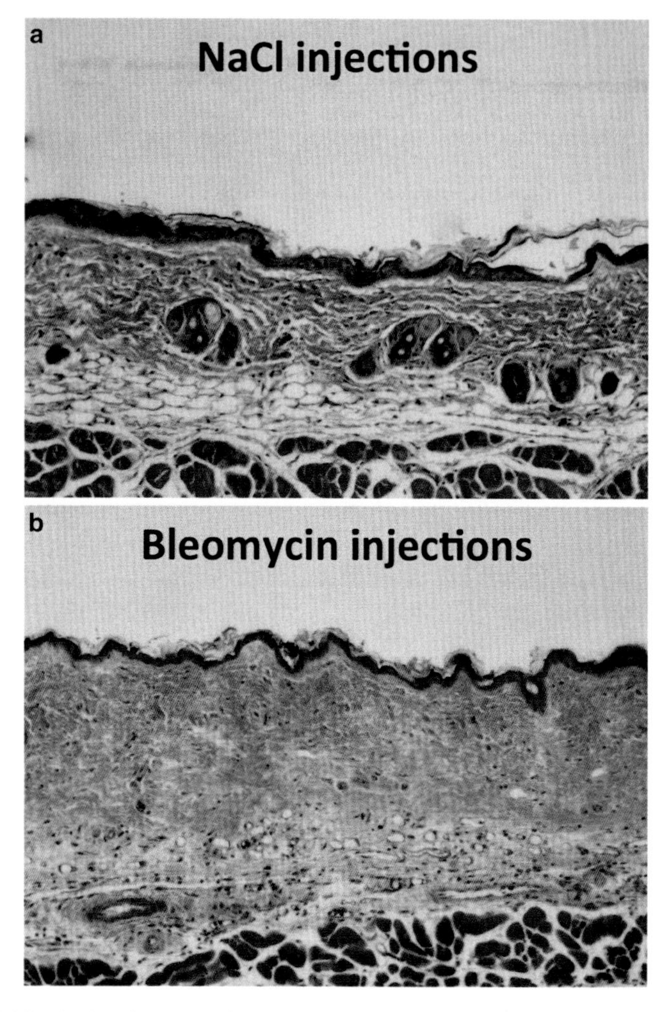

Fig. 2 Histological features of bleomycin-induced skin fibrosis compared to mice injected with sodium chloride. (**a**, **b**) Representative sections stained by hematoxylin–eosin at 100-fold magnification. Increased dermal thickness is observed in mice injected subcutaneously with bleomycin for 3 weeks (**b**) compared to mice injected with sodium chloride within the same time period (**a**). In addition, thickened collagen bundles and cellular infiltrates are observed in mice treated with bleomycin (**b**)

The bleomycin model may respond to treatment with anti-inflammatory/immunosuppressive therapies that are not effective in the human disease. A major reason for this discrepancy might be that patients are usually seen at the stage of established fibrosis, in which meaningful inflammatory infiltrates are often no longer detectable.

6. To prove the efficacy of anti-fibrotic agents in different stages of SSc, anti-fibrotic effects observed in the bleomycin model should be confirmed in non-inflammatory models of SSc.

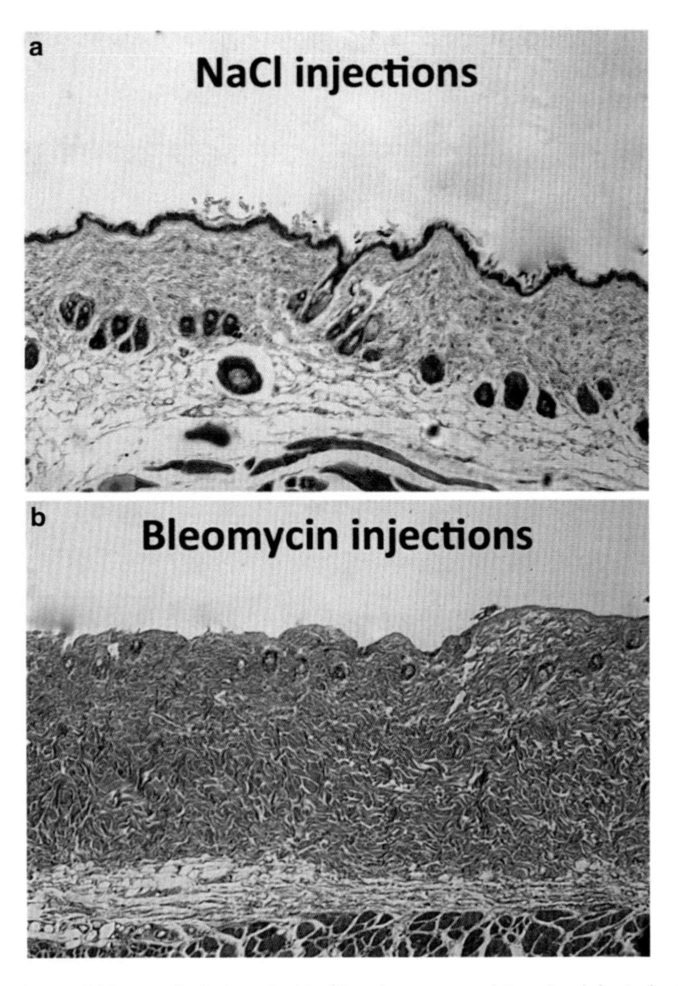

Fig. 3 Histological features of bleomycin-induced skin fibrosis compared to mice injected with sodium chloride. Representative sections stained by trichrome staining at 100-fold magnification. Increased accumulation of collagen is observed in mice subjected to bleomycin injections (**b**) compared to mice injected with sodium chloride (**a**)

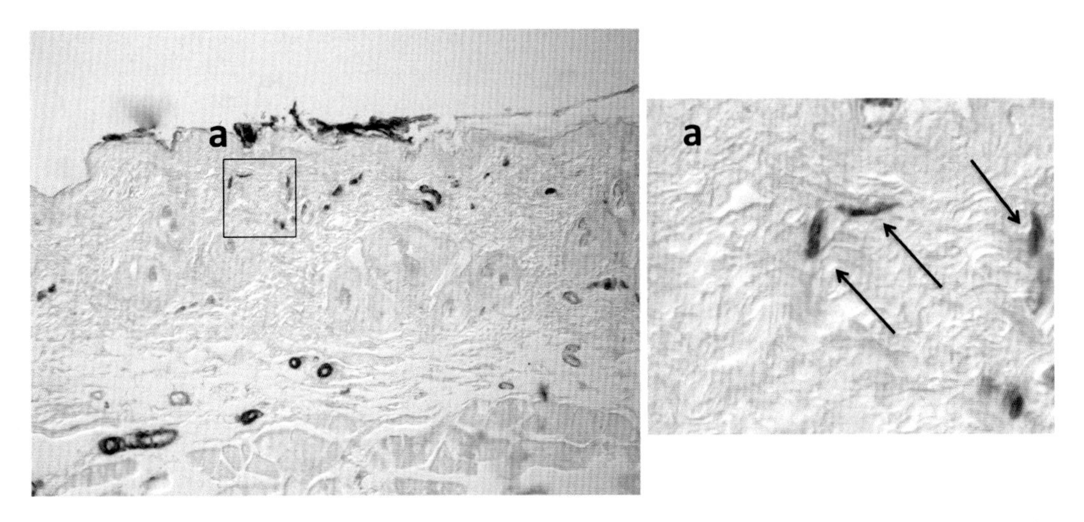

Fig. 4 Representative picture showing immunohistochemistry staining for α-SMA. Myofibroblasts, corresponding to α-SMA positive cells with a spindle-shaped morphology, are pointed by *arrows*

References

1. Gabrielli A, Avvedimento EV, Krieg T (2009) Scleroderma. N Engl J Med 360(19): 1989–2003
2. Yamamoto T, Takagawa S, Katayama I, Yamazaki K, Hamazaki Y, Shinkai H et al (1999) Animal model of sclerotic skin. I: local injections of bleomycin induce sclerotic skin mimicking scleroderma. J Invest Dermatol 112(4):456–462
3. Avouac J, Furnrohr BG, Tomcik M, Palumbo K, Zerr P, Horn A et al (2011) Inactivation of the transcription factor STAT-4 prevents inflammation-driven fibrosis in animal models of systemic sclerosis. Arthritis Rheum 63(3):800–809
4. Avouac J, Palumbo K, Tomcik M, Zerr P, Dees C, Horn A et al (2012) Inhibition of activator protein 1 signaling abrogates transforming growth factor beta-mediated activation of fibroblasts and prevents experimental fibrosis. Arthritis Rheum 64(5):1642–1652
5. Avouac J, Elhai M, Tomcik M, Ruiz B, Friese M, Piedavent M et al (2013) Critical role of the adhesion receptor DNAX accessory molecule-1 (DNAM-1) in the development of inflammation-driven dermal fibrosis in a mouse model of systemic sclerosis. Ann Rheum Dis 72(6):1089–1098
6. Kitaba S, Murota H, Terao M, Azukizawa H, Terabe F, Shima Y et al (2012) Blockade of interleukin-6 receptor alleviates disease in mouse model of scleroderma. Am J Pathol 180(1):165–176
7. Wei J, Zhu H, Komura K, Lord G, Tomcik M, Wang W et al (2014) A synthetic PPAR-gamma agonist triterpenoid ameliorates experimental fibrosis: PPAR-gamma-independent suppression of fibrotic responses. Ann Rheum Dis 73(2):446–454
8. Beyer C, Schett G, Distler O, Distler JH (2010) Animal models of systemic sclerosis: prospects and limitations. Arthritis Rheum 62(10):2831–2844
9. Ishikawa H, Takeda K, Okamoto A, Matsuo S, Isobe K (2009) Induction of autoimmunity in a bleomycin-induced murine model of experimental systemic sclerosis: an important role for CD4+ T cells. J Invest Dermatol 129(7):1688–1695
10. Yoshizaki A, Iwata Y, Komura K, Ogawa F, Hara T, Muroi E et al (2008) CD19 regulates skin and lung fibrosis via Toll-like receptor signaling in a model of bleomycin-induced scleroderma. Am J Pathol 172(6):1650–1663
11. Yamamoto T, Nishioka K (2005) Cellular and molecular mechanisms of bleomycin-induced murine scleroderma: current update and future perspective. Exp Dermatol 14(2):81–95
12. Oi M, Yamamoto T, Nishioka K (2004) Increased expression of TGF-beta1 in the sclerotic skin in bleomycin-'susceptible' mouse strains. J Med Dent Sci 51(1):7–17
13. Takagawa S, Lakos G, Mori Y, Yamamoto T, Nishioka K, Varga J (2003) Sustained activation of fibroblast transforming growth factor-beta/Smad signaling in a murine model of scleroderma. J Invest Dermatol 121(1):41–50
14. Yamamoto T, Nishioka K (2002) Animal model of sclerotic skin. V: increased expression of alpha-smooth muscle actin in fibroblastic cells in bleomycin-induced scleroderma. Clin Immunol 102(1):77–83
15. Yamamoto T, Kuroda M, Nishioka K (2000) Animal model of sclerotic skin. III: histopathological comparison of bleomycin-induced scleroderma in various mice strains. Arch Dermatol Res 292(11):535–541
16. Woessner JF Jr (1961) The determination of hydroxyproline in tissue and protein samples containing small proportions of this imino acid. Arch Biochem Biophys 93:440–447

Chapter 12

Screening for Novel Serum Biomarker for Monitoring Disease Activity in Rheumatoid Arthritis Using iTRAQ Technology-Based Quantitative Proteomic Approach

Satoshi Serada and Tetsuji Naka

Abstract

Useful biomarkers, which enable the prediction of drug susceptibility, identification of side effects, and/or evaluation of disease activity during drug treatment, are urgently needed to select adequate drugs for patients. Gene mutation status, protein expression levels in a biopsy, and serum proteins are often used as biomarkers. One of the methods to screen for protein biomarkers involves quantitative proteomic approaches using mass spectrometry. Owing to the development of quantitative proteomic approaches, the efficiency of identifying novel biomarkers from clinical samples has improved. In particular, isobaric tag for relative and absolute quantitation technology, which enables relative comparative analysis of up to eight samples, enables high-throughput analysis of screening for biomarkers at the protein level. Here, we describe the identification of a novel biomarker, which is useful for the evaluation of disease activity in patients with rheumatoid arthritis who were treated with anti-TNF-α therapy.

Key words Proteomics, iTRAQ, Biomarker, RA, Disease activity marker

1 Introduction

Classical proteomic approaches are mainly based on two-dimensional gel electrophoretic separation of proteins by isoelectric point and molecular weight on a polyacrylamide gel combined with the identification of separated protein spots by mass spectrometric analysis [1, 2]. Two-dimensional gel electrophoresis (2-DE) has the advantage of separating proteins with posttranslational modifications, such as phosphorylation, into distinct spots owing to the alternation of charge. However, regarding the separation of proteins using 2-DE, several limitations were observed. It is difficult to separate high-molecular-weight proteins by isoelectric focusing and membrane proteins because of poor solubility and low expression. When comparing multiple proteome samples, many gels must be used in one experiment, and it is not an easy task to run 2-DE with high reproducibility.

Shunichi Shiozawa (ed.), *Arthritis Research: Methods and Protocols*, Methods in Molecular Biology, vol. 1142, DOI 10.1007/978-1-4939-0404-4_12, © Springer Science+Business Media New York 2014

Shotgun proteomics, which is a gel-free method, enables the direct analysis of complex peptide mixtures obtained from enzymatic digestion of protein mixtures extracted from various samples. This method rapidly generates a global expression profile of proteins [3]. For the identification of a large number of proteins, it is more effective to perform other chromatographic separation methods before reverse-phase (RP) chromatography, such as strong cation-exchange (SCX) chromatography. This enables a decrease in the peptide complexity in each fraction. Systems comprising 2-D liquid chromatography (LC), tandem mass spectrometry (MS/MS), and protein database searching algorithms are widely used in proteomic analysis. In addition, recent proteomic approaches have had many breakthroughs, and these have remarkably improved the sensitivity, accuracy, and resolution of the mass spectrometer. Thus, shotgun analysis shows advantages over 2-DE gel-based proteomics technologies both for detection sensitivity and proteome coverage [4].

The development of novel reagents for quantitative proteomics, pretreatment methodologies, and software for analyzing data obtained from mass spectrometric analysis has enabled the application of shotgun analysis-based proteomic approaches to quantitative proteomics. Quantitative proteomics using shotgun approaches can be classified into labelled or label-free methods. Several chemical labelling technologies have been established, including 18O labelling [5], isotope-coded affinity tag (ICAT) [6], stable isotope-labelled amino acids in culture (SILAC) [7], and isobaric tag for relative and absolute quantitation (iTRAQ) [8, 9] methods. In the case of label-free quantification approaches, spectral counting or LC/MS quantification of MS/MS precursors has been used [10]. Both labelled and label-free methods have their advantages and limitations. Chemical labelling enables better quantification precision; however, it has the limitation of potentially incomplete peptide labelling and unintended side reactions. Label-free methods are simpler; however, quantification outcomes are less precise compared with those for labelled methods. Using iTRAQ analysis, one can detect differentially expressed proteins in four to eight samples in a single mass spectrometric analysis.

The iTRAQ reagent 4plex consists of an N-methylpiperazine quantification group, a carbonyl stable-isotope mass balance group, and a hydroxylsuccinimide ester group that reacts with primary amines on the N-termini of peptides and also the side chains of lysine. Peptide quantification using iTRAQ analysis is based on the relative abundance of four quantification reporter ions (m/z 114, 115, 116, and 117), which are generated via MS/MS fragmentation of iTRAQ-labelled peptide mixtures. iTRAQ technology has several advantages. First, iTRAQ reagents can label nearly all peptides, providing higher peptide coverage and identification of protein sequences for quantitative peptide information.

Second, because all four iTRAQ-labelled samples are combined and represent the same m/z in the MS spectra, each MS spectrum shows the accumulated peak derived from four samples, and the high-quality b-series and y-series ions are generated in the MS/MS spectra, resulting in better ion statistics for a more accurate identification of peptides. Third, the iTRAQ approach is a postlabelling method; labelling is performed at the peptide level after the proteins extracted from the samples were enzymatically digested to peptides. Therefore, the iTRAQ approach has the advantage of being able to analyze various samples, including tissue and body fluids such as serum and plasma. The SILAC approach, a prelabel method, is suitable for analyzing cultured cells, because labelling is performed during cell culture. However, unlike iTRAQ, using this technology to analyze tissue or body fluid samples is not easy.

iTRAQ analysis is applicable for identifying biomarkers by comparing proteomes of tissues or body fluids such as sera obtained from patients [11–13]. Regarding autoimmune disease, iTRAQ analysis has been performed against sera obtained from patients with rheumatoid arthritis (RA) who were treated with biological therapeutics [14, 15].

Here, we present an iTRAQ-based shotgun serum proteomics method that we have used to screen for serum biomarker proteins. Recent development of biological therapies has led to a marked improvement in the treatment of RA and other autoimmune diseases. Because conventional inflammatory biomarkers, including C-reactive protein (CRP), do not reflect disease activity in patients treated with anti-IL-6 biological therapy, new biomarkers are required for the adequate evaluation of therapeutic efficacy. To screen for novel biomarkers that would be highly useful for monitoring the disease activity of patients with RA during therapy, we conducted a proteomics study of sera from patients with RA before and after anti-TNF-α therapy and pooled sera from healthy control (*see* Fig. 1). Highly abundant serum proteins were depleted using a MARS Hu-14 multiple affinity removal LC column, and lesser abundant proteins were first digested with trypsin. Three iTRAQ reagents were used to label proteins in sera from patients with RA before and after anti-TNF-α therapy and proteins in pooled sera of healthy controls. Equal amounts of the labelled peptides were mixed and separated using SCX-HPLC and subsequently analyzed using nano-LC-MS/MS. Bioinformatic analysis of the MS/MS spectra demonstrated a more than 1.5-fold increase in the level of 31 proteins in the sera of patients with RA at baseline compared with that after treatment [16]. Using iTRAQ analysis of the proteome of sera from patients with RA before and after anti-TNF-α therapy, we found a positive correlation between the serum levels of leucine-rich α-2 glycoprotein (LRG) and disease activity in RA (*see* Fig. 2) [16].

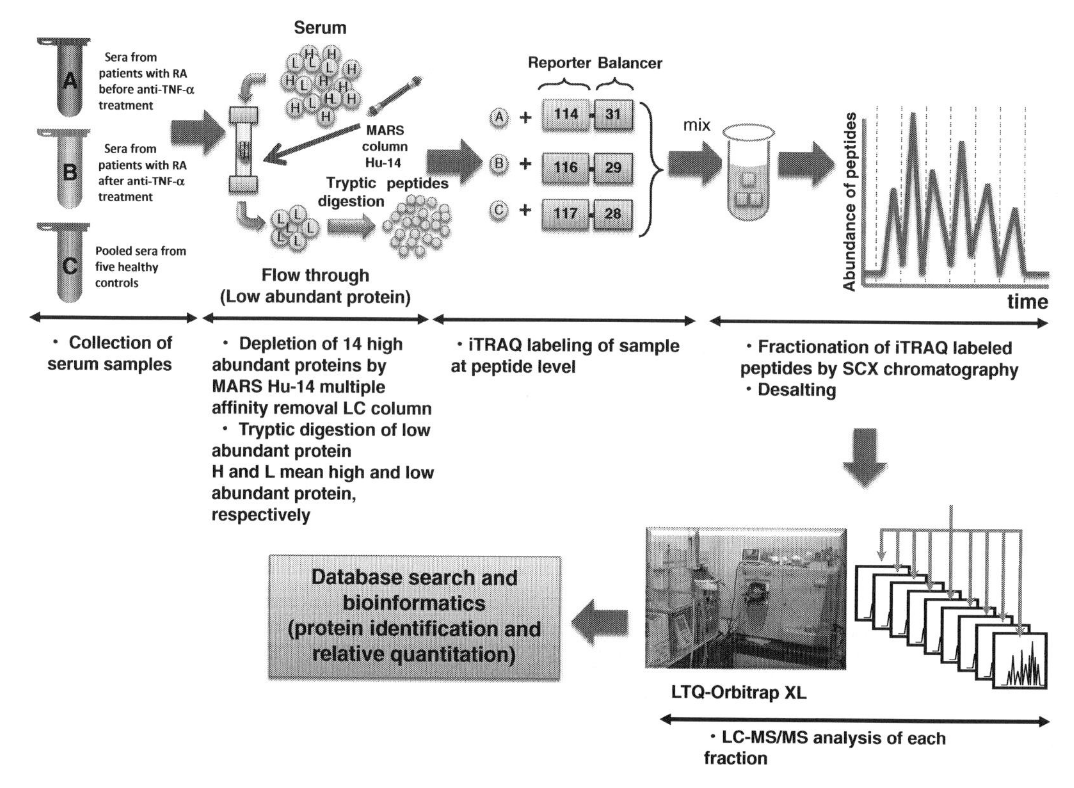

Fig. 1 iTRAQ-based shotgun serum proteomics work flow. Sera are collected from patients with RA before and after anti-TNF-α treatment. Sera are also collected from five healthy controls and pooled. Diluted sera are injected into the MARS Hu-14 multiple affinity removal LC column. Flow-through fractions (containing lesser abundant proteins) are collected and reduced with TCEP and alkylated with MMTS followed by digestion with trypsin. Pretreated serum samples from patients with RA before and after treatment are labelled with the iTRAQ reagents 114 and 116, respectively. A pretreated serum sample from a pooled healthy control sample is labelled with the iTRAQ reagent 117. iTRAQ reagents are composed of N-methylpiperazine quantification group, reporter (*m*/*z* 114, 116, and 117) and balancer (*m*/*z* 31, 29, 28). The labelled peptides are combined and separated using SCX-HPLC and desalted with C18 beads. The labelled peptides are separated using reverse-phase HPLC and analyzed on a tandem mass spectrometer. In each iTRAQ reagent, sum of reporter and balancer are set to be a same *m*/*z* of 145. By this devise, the molecular weight of each iTRAQ reagent-labelled peptides shows the same *m*/*z* in MS spectrum. Peptide quantification is based on the relative abundance of the four quantification reporter ions (*m*/*z* 114, 116, and 117) generated through MS/MS fragmentation of iTRAQ-labelled peptide mixtures as shown in Fig. 2. Database searching and bioinformatics procedures are performed to obtain protein identification and quantification

2 Materials

2.1 Preparation of Human Serum and Depletion of Highly Abundant Serum Proteins

1. BD vacutainer tubes: BD (Toronto, Ontario, Canada).
2. MARS Hu-14 multiple affinity removal LC column (4.6 × 100 mm) and MARS Column Reagent Starter Kit [containing buffer A, buffer B, spin filters (0.22 μm), and concentrators (5 kDa MWCO)]: Agilent Technologies (Santa Clara, CA).
3. Appropriate LC Systems for the MARS Hu-14 multiple affinity removal LC column.

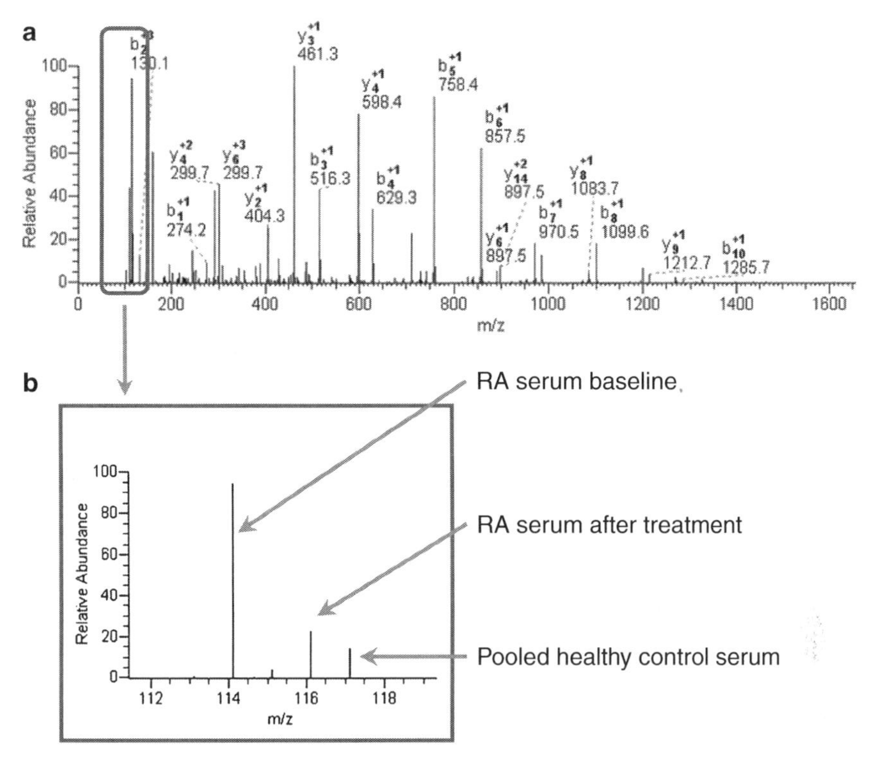

Fig. 2 MS/MS spectrum of a leucine-rich α-2 glycoprotein (LRG) peptide. Elevated level of serum LRG is found in sera from a patient with RA at baseline (high disease activity stage) in comparison with that after treatment (low disease activity stage) and that in a pooled healthy control serum. (**a**) To identify the protein, information regarding the peptide sequence is deduced from the MS/MS spectrum on the basis of the continuous series of N-terminal (b-series) and C-terminal (y-series) ions. (**b**) The peak areas of iTRAQ reporter ions, *m/z* 114 through 117, are used to determine the relative abundance of the peptide

4. DC Protein Assay Kit: Bio-Rad Laboratories (Hercules, CA).
5. 2-D Clean-Up Kit: GE Healthcare BioSciences (Little Chalfont, UK).
6. 1 M TEAB: Sigma Chemical Co. (St. Louis, MO).

2.2 iTRAQ Labelling

1. iTRAQ Reagent 4Plex Multiplex Kit (AB Sciex, Foster City, CA): The iTRAQ Reagent 8Plex Multiplex Kit is also available for the analysis of 5–8 samples.
2. Trypsin with $CaCl_2$ (AB Sciex).
3. Sep-Pak C18 cartridge: Waters Corporation (Milford, MA, USA).

2.3 SCX Fractionation

1. SCX column: ZORBAX 300-SCX 2.1 × 150 mm 5 μm P.N.883700-704.
2. Appropriate HPLC systems (such as Agilent1200) for the SCX column.
3. Peptide clean-up C18 spin tubes: Agilent Technologies.

2.4 Mass Spectrometric Analysis

1. LTQ-OrbitrapXL: Thermo Fisher Scientific (Waltham, MA).
2. Paradigm MG2 pump: Michrom Bioresources (Auburn, CA).
3. Autosampler HTC PAL: CTC Analytics (Zwingen, Switzerland).
4. MagicC18AQ column: 100 μm × 150 mm, 3.0-μm particle size, 300 Å (Michrom Bioresources).
5. Formic acid: Wako Pure Chemical Industries Ltd. (Tokyo, Japan).
6. Acetonitrile (LC/MS grade) and distilled water (LC/MS grade): Kanto Kagaku (Tokyo, Japan).

3 Methods

Perform all procedures at room temperature unless specified otherwise.

3.1 Preparation of Human Serum and Depletion of Highly Abundant Serum Proteins

1. For serum samples, collect 5 ml of blood by venipuncture into the BD vacutainer serum separation tubes (SST), allow to clot at room temperature for 30 min, and centrifuge at 1,300 × *g* for 10 min. Remove serum, immediately divide into 100 μl and 1 ml aliquots, and store at −80 °C until use (*see* **Note 1**).
2. Set up buffer A and buffer B as the only mobile phases (buffer A and buffer B are included in the MARS Column Reagent Starter Kit). Flush the LC system with buffer A and buffer B at a flow rate of 1.0 ml/min for 10 min without a column (*see* **Note 2**).
3. Set up the LC timetable (*see* Table 1), and run two method blanks by injecting 200 μl of buffer A without a column. Ensure proper sample loop size in the autosamplers.

Table 1
LC timetable for MARS Hu-14 multiple affinity removal LC column

Solvent A: Buffer A				
Solvent B: Buffer B				
LC timetable				
	Time	%B	Flow rate (ml/min)	Max pressure (bar)
1	0.00	0.00	0.125	60
2	18.00	0.00	0.125	60
3	18.01	0.00	1.000	60
4	20.00	0.00	1.000	60
5	20.01	100.00	1.000	60
6	27.00	100.00	1.000	60
7	27.01	0.00	1.000	60
8	38.00	0.00	1.000	60

4. Attach a column, and equilibrate it with buffer A for 4 min at a flow rate of 1 ml/min at room temperature.
5. Perform a fourfold dilution of human serum/plasma with buffer A. For sample dilution, add protease inhibitors to buffer A to help prevent protein degradation.
6. Remove particulates using a 0.22-μm spin filter for 1 min at 16,000 × *g*.
7. Inject 160 μl of the diluted serum/plasma at a flow rate of 0.125 ml/min.
8. Collect the flow-through fraction (which appears between 11 and 15 min), and store collected fractions at −20 °C if they are not analyzed immediately.
9. Elute bound proteins from the column with buffer B (elution buffer) at a flow rate of 1 ml/min for 7 min.
10. Regenerate the column by equilibrating it with buffer A for 11 min at a flow rate of 1 ml/min.
11. Store the column after equilibrating it with buffer A at 2–8 °C in a refrigerator.
12. Concentrate collected flow-through fractions using concentrators (5 kDa) until sample volumes are enriched to approximately 100 μl.
13. Quantitate protein concentrations using the DC Protein Assay Kit.
14. Precipitate and desalt each protein (100 μg) using a 2-D Clean-Up Kit.
15. Resolve precipitated proteins using 10 μl of 7 M urea, 2 M thiourea, and 1 % CHAPS buffer.

3.2 iTRAQ Labelling

1. Dilute the samples using the 11 μl of 1 M TEAB (pH 8.5) solution.
2. Reduce proteins using 2 μl of reducing reagent [50 mM Tris-(2-carboxyethyl) phosphine (TCEP)], and mix samples by vortexing and spinning down.
3. Perform the reduction reaction at room temperature for 1 h.
4. Alkylate reduced proteins by adding 1 μl of 200 mM methylmethanethiosulfonate (MMTS)/isopropanol, and mix samples by vortexing and spinning down.
5. Perform alkylation at room temperature for 10 min.
6. Before adding trypsin, add 366 μl of the 50 mM TEAB (pH 8.5) solution to samples to dilute reagents that would affect the activity of trypsin (such as urea, thiourea, and CHAPS).
7. Add 10 μl of 1 mg/ml trypsin with $CaCl_2$ to each sample. Mix samples by gently vortexing.
8. Digest proteins with trypsin overnight at 37 °C.

9. After tryptic digestion, concentrate samples using SpeedVac until sample volumes decrease to 30 μl.
10. For iTRAQ 4plex labelling, mix each iTRAQ reagent (4plex) by vortexing and spinning down (*see* **Note 3**).
11. Add 70 μl of ethanol to each tube of iTRAQ reagents and mix by vortexing and spinning down.
12. Add the diluted iTRAQ reagents to the samples and mix by vortexing and spinning down.
13. Perform the iTRAQ labelling reaction at room temperature for 1 h.
14. After iTRAQ labelling, stop the labelling reaction by adding 1.0 ml of distilled water.
15. In a 50-ml tube, add 6.0 ml of distilled water and mix four of these diluted iTRAQ-labelled samples in this tube. As a result, the concentration of the organic solvent should be below 3 %.
16. Decrease the pH of the sample to 2.5–3.0 by adding 10 % TFA (*see* **Note 4**).
17. Desalt peptides using a Sep-Pak C18 cartridge. Using a 2-ml all-plastic syringe, add 2 ml of 10 % distilled water/90 % acetonitrile to the cartridge.
18. Using a 2-ml all-plastic syringe, and add 2 ml of 98 % distilled water/2 % acetonitrile to the cartridge.
19. Using a 20-ml all-plastic syringe, add the sample to the cartridge, collect the flow-through, and add the flow-through to the cartridge again.
20. Wash the cartridge with 5 ml of 98 % distilled water/2 % acetonitrile using a 5-ml all-plastic syringe.
21. Elute the peptide from the cartridge using a 1 ml all-plastic syringe containing 10 % distilled water/90 % acetonitrile.
22. Dry the eluted peptides using SpeedVac.

3.3 SCX Fractionation

1. Dissolve the dried sample in 50 μl of buffer A [25 % acetonitrile (v/v); 10 mM potassium phosphate, pH 3.0].
2. Adjust the pH to 3.0 with 0.1 N NaOH or 1 M phosphate. Add buffer A [25 % acetonitrile (v/v); 10 mM potassium phosphate, pH 3.0] up to a volume of 100 μl.
3. Centrifuge at 18,100 × *g* for 5 min at 4 °C. Transfer supernatant into the vial for the autosampler.
4. Set up the HPLC timetable (*see* Table 2).
5. Inject the samples into the SCX column. Collect each fraction into the 96-well microplate at 2-min intervals (0.4 ml/well).
6. After SCX chromatography, transfer each fraction into the 1.5-ml microtube and lyophilize. Collect fractions 4–35.

Table 2
LC timetable for SCX chromatography separating iTRAQ-labelled peptides

Solvent A: 10 mM KH_2PO_4, 25 % CH_3CN, pH3				
Solvent B: 10 mM KH_2PO_4, 25 % CH_3CN, 1 M KCl, pH3				
LC timetable				
	Time	%B	Flow rate (ml/min)	Max pressure (bar)
1	0	0.0	0.200	400
2	10	0.0	0.200	400
3	40	8.8	0.200	400
4	60	35.0	0.200	400
5	61	100.0	0.200	400
6	65	100.0	0.200	400
7	66	0.0	0.200	400
8	80	0.0	0.200	400

7. Desalt the samples using the peptide clean-up C18 spin tubes.

   Resolve lyophilized samples using 100 μl of 0.5 % TFA (v/v)/5 % acetonitrile (v/v)/95 % distilled water (v/v). Some low-peptide-containing fractions may be pooled. Fractions 4–7, 8–11, 12 and 13, and 14 and 15 may be pooled; this will result in a total of 24 samples.

8. Add 200 μl of 50 % acetonitrile (v/v) to the peptide clean-up C18 spin tubes and centrifuge at 1,500 × *g* for 1 min at room temperature. Repeat this step again.

9. Add 200 μl of 0.5 % TFA (v/v)/5 % acetonitrile (v/v)/95 % distilled water (v/v) to the peptide clean-up C18 spin tubes and centrifuge at 1,500 × *g* for 1 min at room temperature. Repeat this step again.

10. Add samples to the peptide clean-up C18 spin tubes, and place the column onto the original sample tube and centrifuge at 1,500 × *g* for 1 min at room temperature. Repeat this step again.

11. To wash the resin, place the column onto the 2-ml collection tube, add 200 μl of 0.5 % (v/v) TFA/5 %/acetonitrile/95 % (v/v)/distilled water to the peptide clean-up C18 spin tubes, and centrifuge at 1,500 × *g* for 1 min at room temperature. Repeat this step again five times.

12. To elute the peptide from the resin, place the column onto a new 1.5-ml microtube, add 35 μl of 70 % acetonitrile (v/v), and centrifuge at 1,500 × *g* for 1 min at room temperature. Repeat this step again.

13. Dry the peptide using SpeedVac.
14. Dissolve the dried fractions in 30 μl of 0.1 % TFA (v/v) and 2 % acetonitrile (v/v)/98 % (v/v) distilled water and centrifuge at 16,100 ×*g* for 5 min at 4 °C. Transfer supernatant into the vial for the autosampler for subsequent LC-MS/MS analysis.

3.4 Mass Spectrometric Analysis

1. Mass spectrometric analysis: NanoLC-MS/MS analyses were performed on an LTQ-OrbitrapXL (Thermo Fisher Scientific) equipped with a nano-ESI source and coupled to a Paradigm MG2 pump (Michrom Bioresources) and the autosampler (HTC PAL; CTC Analytics). Peptide mixtures were separated on a MagicC18AQ column (100 μm × 150 mm, 3.0 μm particle size, 300 Å, Michrom Bioresources) with a flow rate of 500 nl/min. A linear gradient of 5–30 % B in 80 min, 30–95 % B in 10 min, and 95 % B in 4 min, which was finally decreased to 5 % B, was used (A = 0.1 % formic acid in 2 % acetonitrile, B = 0.1 % formic acid in 90 % acetonitrile). Up to three HCD spectra were acquired in a data-dependent acquisition mode following each full scan (*m/z*, 350–1,500).
2. iTRAQ data analysis: Protein identification and quantification for iTRAQ analysis were performed using Proteome Discoverer version 1.3 (Thermo Fisher Scientific) against the Swiss Prot human protein database. Search parameters for peptide and MS/MS mass tolerance were 10 ppm and 0.8 Da, respectively, with the allowance of two missed cleavages made from the trypsin digest. Search parameters included iTRAQ labelling at the N-terminus and lysine residues and cysteine modification by MMTS. The false discovery rate was confirmed to be less than 5 %. Relative protein abundances were calculated using the ratio of iTRAQ reporter ions in the MS/MS scan.
3. For validation purposes, it is very important to verify the iTRAQ quantification results using western blotting, immunohistochemistry, or other biological methods.

4 Notes

1. Proposals for research using human blood cells should undergo a process of ethical review and approval.
2. About approximately 20 types of highly abundant proteins account for the 99 % of proteins in serum or plasma. It is difficult to analyze a serum or a plasma proteome without pretreatment. To increase the number of proteins identified, one of the effective pretreatment techniques is to deplete majorly expressing proteins from serum or plasma using an antibody-immobilized column. Several columns are now commercially available. The MARS Hu-14 multiple affinity removal LC column is designed

to remove 14 highly abundant proteins from human biological fluids such as plasma, serum, and cerebral spinal fluid (CSF). This column enables the removal of albumin, IgG, antitrypsin, IgA, transferrin, haptoglobin, fibrinogen, α-2-macroglobulin, α-1-acid glycoprotein, IgM, apolipoprotein AI, apolipoprotein AII, complement C3, and transthyretin. The targeted highly abundant proteins are simultaneously removed when crude biological samples are passed through the column. Specific removal of 14 highly abundant proteins depletes approximately 94 % of the total protein mass from human serum or plasma. The lesser abundant proteins in the flow-through fractions can be analyzed by quantitative proteomic approaches such as iTRAQ analysis.

3. When the iTRAQ reagent 8plex was used, add 50 μl of isopropanol. Mix by vortexing, spin down, and add the diluted iTRAQ 8plex reagent to the samples. Mix by vortexing, and spin down. Perform the iTRAQ labelling reaction at room temperature for 1 h. Stop the iTRAQ labelling reaction by adding 1.0 ml of distilled water to each sample. In a 50-ml tube, add 10.0 ml of distilled water and mix four of these diluted iTRAQ-labelled samples in this tube. As a result, the concentration of the organic solvent should be below 3 %.
4. Before mass spectrometric analysis, remove excess unreacted iTRAQ reagents and decrease the peptide complexity of peptides using SCX chromatography. This is highly effective and increases protein identification and peptide coverage by mass spectrometric analysis. To enrich iTRAQ-labelled peptides using the Sep-Pak C18 cartridges, it is necessary to dilute the concentration of the organic solvent and adjust the pH.

References

1. Lilley KS, Rassaq A, Dupree P (2002) Two-dimensional gel electrophoresis: recent advances in sample preparation, detection and quantitation. Curr Opin Chem Biol 6:46–50
2. Gharbi S, Gaffney P, Yang A, Zvelebil MJ, Cramer R, Waterfield MD, Timms JF (2002) Evaluation of two-dimensional differential gel electrophoresis for proteomic expression analysis of a model breast cancer cell system. Mol Cell Proteomics 1:91–98
3. Wu CC, MacCoss MJ (2002) Shotgun proteomics: tools for the analysis of complex biological systems. Curr Opin Mol Ther 4:242–250
4. McDonald WH, Yates JR 3rd (2002) Shotgun proteomics and biomarker discovery. Dis Markers 18:99–105
5. Wu WW, Wang G, Yu MJ, Knepper MA, Shen RF (2007) Identification and quantification of basic and acidic proteins using solution-based two-dimensional protein fractionation and label-free or (18) o-labeling mass spectrometry. J Proteome Res 6:2447–2459
6. Shiio Y, Aebersold R (2006) Quantitative proteome analysis using isotope-coded affinity tags and mass spectrometry. Nat Protoc 1:139–145
7. Mann M (2006) Functional and quantitative proteomics using SILAC. Nat Rev Mol Cell Biol 7:952–958
8. Ross PL, Huang YN, Marchese JN, Williamson B, Parker K, Hattan S, Khainovski N, Pillai S, Dey S, Daniels S, Purkayastha S, Juhasz P, Martin S, Bartlet-Jones M, He F, Jacobson A, Pappin DJ (2004) Multiplexed protein quantitation in Saccharomyces cerevisiae using amine-reactive isobaric tagging reagents. Mol Cell Proteomics 3:1154–1169
9. Wiese S, Reidegeld KA, Meyer HE, Warscheid B (2007) Protein labeling by iTRAQ: a new tool for quantitative mass spectrometry in proteome research. Proteomics 7:1004

10. Zhang B, VerBerkmoes NC, Langston MA, Uberbacher E, Hettich RL, Samatova NF (2006) Detecting differential and correlated protein expression in label-free shotgun proteomics. J Proteome Res 5:2909–2918
11. Garbis SD, Tyritzis SI, Roumeliotis T, Zerefos P, Giannopoulou EG, Vlahou A, Kossida S, Diaz J, Vourekas S, Tamvakopoulos C, Pavlakis K, Sanoudou D, Constantinides CA (2008) Search for potential markers for prostate cancer diagnosis, prognosis and treatment in clinical tissue specimens using amine-specific isobaric tagging (iTRAQ) with two-dimensional liquid chromatography and tandem mass spectrometry. J Proteome Res 7:3146–3158
12. Waldemarson S, Krogh M, Alaiya A, Kirik U, Schedvins K, Auer G, Hansson KM, Ossola R, Aebersold R, Lee H, Malmström J, James P (2012) Protein expression changes in ovarian cancer during the transition from benign to malignant. J Proteome Res 11:2876–2889
13. Muraoka S, Kume H, Watanabe S, Adachi J, Kuwano M, Sato M, Kawasaki N, Kodera Y, Ishitobi M, Inaji H, Miyamoto Y, Kato K, Tomonaga T (2012) Strategy for SRM-based verification of biomarker candidates discovered by iTRAQ method in limited breast cancer tissue samples. J Proteome Res 11:4201–4210
14. Dwivedi RC, Dhindsa N, Krokhin OV, Cortens J, Wilkins JA, El-Gabalawy HS (2009) The effects of infliximab therapy on the serum proteome of rheumatoid arthritis patients. Arthritis Res Ther 11:R32
15. Ortea I, Roschitzki B, Ovalles JG, Longo JL, de la Torre I, González I, Gómez-Reino JJ, González A (2012) Discovery of serum proteomic biomarkers for prediction of response to infliximab (a monoclonal anti-TNF antibody) treatment in rheumatoid arthritis: an exploratory analysis. J Proteomics 21:372–382
16. Serada S, Fujimoto M, Ogata A, Terabe F, Hirano T, Iijima H, Shinzaki S, Nishikawa T, Ohkawara T, Iwahori K, Ohguro N, Kishimoto T, Naka T (2010) iTRAQ-based proteomic identification of leucine-rich alpha-2 glycoprotein as a novel inflammatory biomarker in autoimmune diseases. Ann Rheum Dis 69: 770–774

Chapter 13

Genome-Wide Genetic Study in Autoimmune Disease-Prone Mice

Masaomi Obata, Mareki Ohtsuji, Yukiyasu Iida, Toshikazu Shirai, Sachiko Hirose, and Hioroyuki Nishimura

Abstract

Mouse models of autoimmune diseases provide invaluable insights into the cellular and molecular bases of autoimmunity. Genetic linkage studies focusing on their abnormal quantitative phenotypes in relation to the loss of self-tolerance will lead to the identification of polymorphic genes that play pivotal roles in the genetic predisposition to autoimmunity. In this chapter, we first overview the basic concepts in the statistical genetics and then provide guides to genotyping microsatellite DNA markers and to quantitative trait loci mapping using a MAPMAKER program.

Key words Quantitative trait loci (QTLs), QTL mapping, Microsatellite markers, Genetic linkage, Disease susceptibility, MAPMAKER program

1 Introduction

1.1 Threshold Liability Model of Complex Diseases

Complex diseases are caused by a multitude of genetic and environmental factors. The best known theory for the polygenic involvement of complex diseases is the threshold liability model proposed by Falconer in 1981 [1]. In this model, the phenotype of an individual is determined by an underlying continuous quantitative variable with a threshold that imposes the disease manifestation. The variable is called *liability* and is assumed to linearly correlate with the total number of disease-susceptibility alleles that an individual inherits. When the liability of an individual is below the threshold, the phenotype of the individual is normal; when it is above the threshold, the individual shows the onset of disease. In reality, susceptibility alleles may contribute differently to the liability (Fig. 1). Some alleles may have either a dominant or a codominant effect on the liability (as shown by *A* and *B*, respectively, in Fig. 1). Others may be dominant suppressor alleles (as shown by *M* in Fig. 1) that

Shunichi Shiozawa (ed.), *Arthritis Research: Methods and Protocols*, Methods in Molecular Biology, vol. 1142, DOI 10.1007/978-1-4939-0404-4_13,

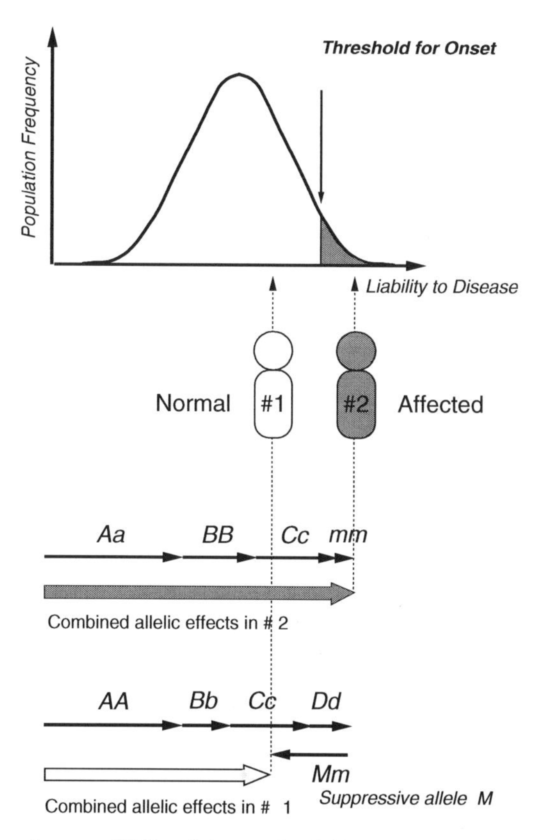

Fig. 1 Effects of susceptibility alleles on the liability to polygenic disease. There are differences in the degree of contribution to the liability among the alleles involved in disease susceptibility. Alleles *A* and *B* are the dominant and codominant alleles contributing to the liability, respectively. The recessive allele *m* makes only a minimum contribution to the liability, while the wild-type allele *M* serves as a dominant suppressor allele exerting a negative effect on the overall liability. While individual #2 develops disease due to having an overall liability higher than the threshold, individual #1 is free from disease due to the negative effect of a dominant suppressor allele *M*

inhibit the onset of disease. There may also be epistatic interactions between the susceptibility alleles of different loci. Despite all these complexities, as long as multiple loci are involved, the allelic effect of each susceptibility gene can be formulated approximately as an additive effect, which is summed up in the total genetic contribution to the disease liability. This model provides the basis for applying quantitative genetics to the study of complex diseases.

Methods of quantitative genetics have been developed and applied to studies on the traits of plants and animals, particularly those of agricultural importance [2]. These quantitative phenotypes are often governed by large numbers of genes, each contributing such a small amount to the phenotype that their individual effects are not detectable by simple *Mendelian* methods. Genes of this nature are termed polygenes, and loci determining the quantitative traits are referred to as quantitative trait loci (QTLs). Quantitative traits are also affected by non-genetic factors such as environmental factors. Therefore, quantitative traits are studied using statistical methods [3]. The availability of multiple genetic markers on the genomic DNA, such as those genotyped by restriction fragment length polymorphism (RFLP) analysis on Southern blots, prompted a genome-wide study to identify QTLs on the chromosomes in the progeny of experimental crosses of animal or plant strains [4]. Computer programs developed for QTL mapping [5] as well as the advent of PCR technology have boosted genetic studies on various quantitative traits, including those relating to disease susceptibilities. In regard to the studies on autoimmune model mice, QTLs regulating the scores for lupus nephritis [6–8], abnormal serum IgG levels [9], ages at onset and titers of autoantibodies [10], frequencies of immune cells with abnormal phenotypes [11, 12], and defective immune tolerance induction [13] have been studied, and the results provided insights into the molecular basis of autoimmune predisposition. The following section provides an overview of the basic concepts of statistical genetics that are required for understanding the method of QTL mapping.

1.2 Maximum Likelihood Method of Genetic Linkage Analysis

Genetic linkage analysis refers to the ordering of genetic loci on a chromosome and to estimating genetic distances among them [14]. When a disease phenotype is linked to a particular genetic marker, the genetic distance between the disease locus and the marker locus on the chromosome is estimated. Among various methods, the *maximum likelihood method* has been widely applied to the studies in human genetic linkage. This method was modified for use in mapping quantitative traits in experimental animals. Some of the basic concepts used in the maximum likelihood method are as follows.

Likelihood is a measure of the plausibility of the observed data, and it is calculated for a particular hypothesis. Likelihood $L(H)$ is defined as the probability with which the observation has occurred under the hypothesis H. Suppose that a disease phenotype in an example of human pedigree (Fig. 2) appears to be associated with a particular marker allele A. There are two mutually exclusive hypotheses: H_1 posits that the disease allele D is linked to the marker allele A, while the alternative hypothesis, H_0 (null hypothesis), asserts that the disease locus and the marker locus are not on the same chromosome. The likelihoods $L(H_1)$ and $L(H_0)$ are calculated for these two hypotheses. When comparing $L(H_1)$ and $L(H_0)$, their absolute values are not generally used. The odds in favor of hypothesis H_1 versus H_0 are expressed by the *likelihood ratio* (R) as follows:

$$R = \frac{L(H_1)}{L(H_0)}$$

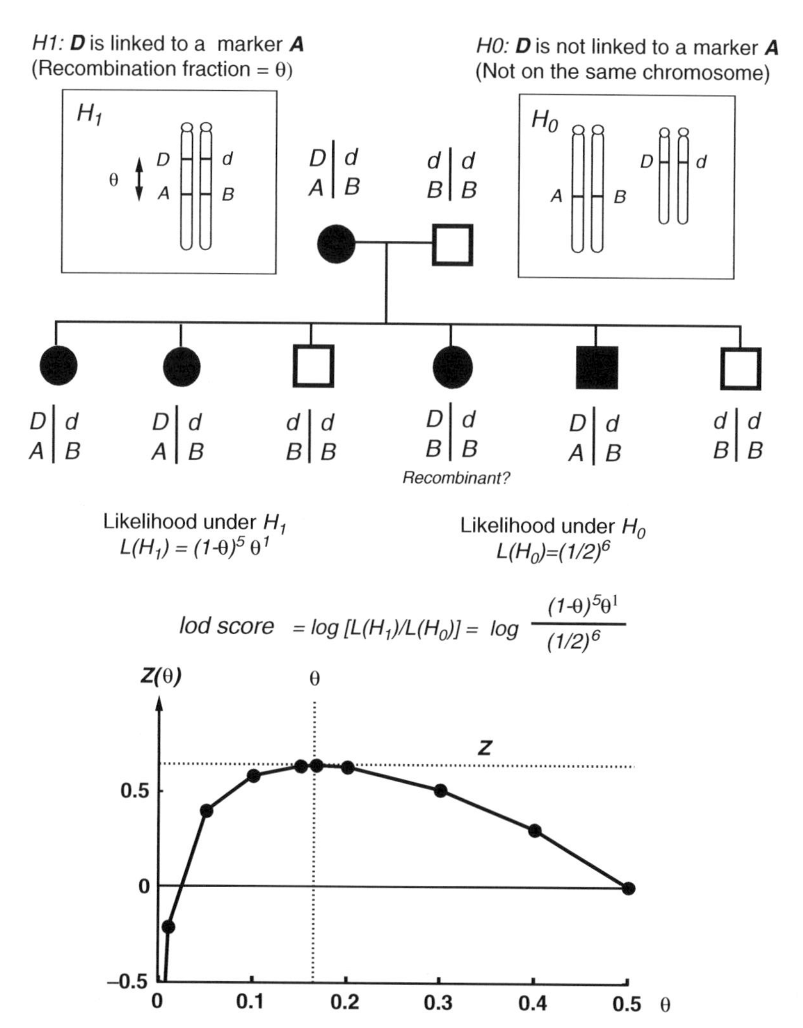

Fig. 2 The lod score method of linkage analysis. Suppose that, in a family with six children, the mother is affected due to the presence of an autosomal dominant disease allele *D* and that four out of the six children are also affected. The disease phenotype will tend to be associated with a marker allele *A* with one exception: two children are healthy and do not have the marker allele *A*. The linkage is evaluated by the lod score as follows. The null hypothesis (H_0) assumes that the disease locus and the marker locus are not on the same chromosome, and the alternative hypothesis (H_1) assumes that *D* is linked to *A* with a recombination fraction θ ($0<\theta<1/2$). The *likelihoods* of the observation are calculated under the two hypotheses. Under H_0, the genotype inherited from the mother is either of non-recombinant type (*DA*, *dB*) or recombinant type (*DB*, *dA*) with an equal probability of 1/2. Therefore, the likelihood is equal to $(1/2)^5 \times (1/2)1 = (1/2)^6$. Conversely under H_1, as recombination between *D* and *A* occurs with a probability of θ, and only one out of six children is the recombinant, the *likelihood* of the observation is formulated as $(1-\theta)^5 \times \theta$. A lod score curve is obtained by plotting [$Z(\theta)$ = logarithm of the likelihood ratio, $L(H_1)/L(H_0)$] at varying θ ($0<\theta<1/2$) values. The lod score takes a maximum at $\theta = 0.167$. This θ value gives a maximum likelihood estimate (MLE) of θ

The *lod score* (logarithm of odds), which is the logarithm of the likelihood ratio, is used as a measure of support for linkage versus no linkage, as follows:

$$\text{Lod score} = \log(R) = \log\left[\frac{L(H_1)}{L(H_0)}\right]$$

In the example of a pedigree shown in Fig. 2, the lod score is expressed as a function of parameter θ (the unknown genetic distance between the disease locus and the marker locus). The *lod score curve* is illustrated on varying θ values ($0 \leqq \theta \leqq 1/2$), and the most plausible θ [*the maximum likelihood estimate*: (MLE) of θ] is estimated by searching for the θ value that gives the highest lod score.

1.3 Method of Mapping Quantitative Trait Loci

The method of QTL mapping was developed by adapting the maximum likelihood method outlined above. For example, a population of experimental animals produced by a backcross of an F1 hybrid strain to one of the parental inbred strains is considered. Suppose that two groups of the population divided by the presence (1) or the absence (0) of an allele M_1 of a genetic marker locus m_1 showed substantial difference with respect to their average values of the quantitative phenotype Φ (Fig. 3a). According to the model in quantitative genetics, Φ_i of an individual i is formulated by the equation

$$\Phi_i = a + b \times g_i + \varepsilon$$

where a is an average Φ value observed in the population without marker allele M_1, g_i is either 0 or 1 according to the presence or the absence of a genetic effect tightly linked to marker allele M_1, and ε is a random normal variable (environmental factor) with mean 0 and variance σ^2. Two hypotheses are considered: H_0 (null hypothesis) and H_1 (linkage). H_0 assumes $b=0$ in the equation, which means that the marker allele M_1 does not affect the phenotypic value. The observed difference of the phenotypic value between the two populations is caused by chance. On the other hand, H_1 assumes $b>0$, which means that the marker allele M_1 affects the phenotypic value. The difference of the average Φ value between the two populations is real. The likelihoods $L(H_0)$ and $L(H_1)$ are calculated. The MLE of a, b, and σ^2 by the regression analysis enables the calculation of the likelihoods, and the significance of the genetic effect of the marker allele M_1 is evaluated by the lod score.

Suppose that a QTL regulating Φ is located on a chromosome and that multiple genetic markers (m_1 to m_n) on the chromosome are analyzed for each individual in the backcross progeny. The lod score is calculated for each marker locus m on the hypothesis H_1, which assumes that the QTL is tightly linked to the marker locus m. If the lod scores are plotted on the genetic map on the chromosome, the marker locus that is most proximal to the real QTL

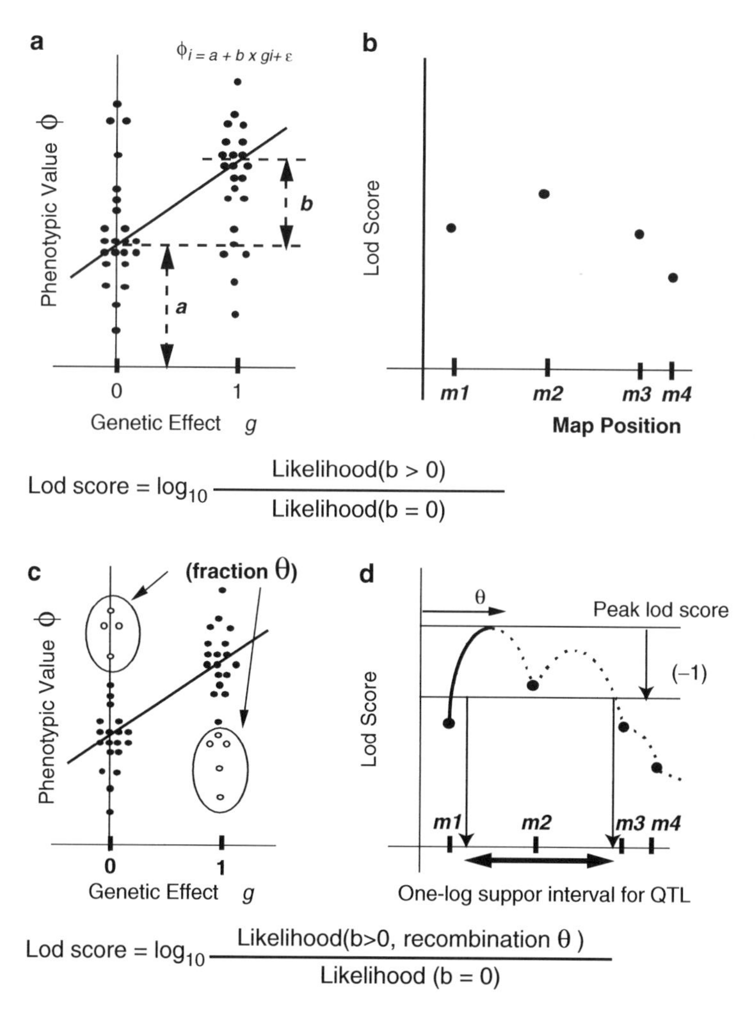

Fig. 3 Method of mapping the quantitative trait locus (QTL). (**a**) Suppose that a population of a progeny produced by backcrossing an F1 hybrid strain with one of the parental strains is grouped into two populations according to the presence (0) or the absence (1) of a marker allele *M*. These two groups show different averages with respect to the phenotypic value Φ. Two hypotheses, H_0 (the null hypothesis) and H_1, are considered. H_0 assumes that the QTL is not linked to the marker allele *M*, and therefore, the observed difference of their mean Φ values is caused by chance. H_1 assumes that the difference is due to the QTL tightly linked to a marker allele *M*. The two populations, positive and negative for the QTL, have different mean Φ values but should have equal deviation around their mean Φ caused by the environmental effect. Maximum likelihood estimates of the factors in the equation $\Phi_i = a + b \times g_i + \varepsilon$ enable the evaluation of lod score for the linkage. (**b**) Suppose that a QTL regulating Φ is located on a chromosome and that multiple genetic markers (m_1 to m_4) on the chromosome are analyzed for each individual in the backcross progeny. The lod score is calculated for each

should show the highest lod score (Fig. 3b). This approach enables a rough estimate of the location of QTL. However, to determine the map position of a QTL more precisely, a parametric approach is required as outlined below.

Suppose that a QTL is on a chromosome and is distant from a marker locus m by a genetic distance θ. In any individual in the backcross progeny, the genotype of the QTL is unknown and may be different from that of marker locus m by a recombination fraction θ. In the two populations of the progeny as grouped by the absence or the presence of a marker allele M, this assumption of the distance between the QTL and the marker locus is in favor of the observation of small populations in Fig. 3c, as depicted by open circles whose Φ values are abnormally high or low as compared with the mean Φ value in each group. Hypothesis H_1 incorporates this assumption, and the corresponding likelihood $L(H)$ is expressed as a function of θ. Similarly, the QTL may be postulated to be located at map position θ within the chromosomal interval between m_n and m_{n+1} on the chromosome. Given that all the marker loci on the chromosome are genotyped for all the individuals in the backcross progeny, the likelihood assuming the presence of QTL on the chromosomal interval is formulated as follows [15]:

$$L(H) = \prod_i \left[G_i(0) L_i(H_0) + G_i(1) L_i(H_1) \right]$$

where $G(0)$ denotes the probability that the QTL genotype of an individual is a type that does not contribute to Φ, and $G(1)$ denotes the probability that the QTL genotype is a type that does contribute to Φ by a factor b. $G(0)$ and $G(1)$ are dependent on the map position of QTL (θ) and the marker genotypes at m_n and m_{n+1} observed in an individual. An algorithm for calculating the likeli-

Fig. 3 (continued) marker locus m under the hypothesis H_1, which assumes that a QTL is tightly linked to the marker locus m. The marker locus that is most proximal to the real QTL should show the highest lod score. (**c**) The parametric method in QTL mapping is explained. Suppose that in each of the two populations, one positive and the other negative for the marker allele M, there are small subsets which show abnormally high or low Φ when compared with the major population in each group. Hypothesis H_1 incorporates the assumption that such small subsets are due to the recombination between the marker locus and QTL, and the corresponding likelihood $L(H_1)$ is expressed as a function of θ. (**d**) Lod scores are also calculated for the map positions located in the intervals of the genetic markers (interval mapping). The peak of the lod curve provides the maximum likelihood estimate of the position of the QTL on the chromosome. The "one-log support interval" is obtained by drawing a horizontal line at 1 unit of lod score below the peak of the lod score curve. The points of intersection of this line with the lod score curve are projected onto the genetic map, which shows the genomic interval in which the QTL should be found with a probability of 90 %

Table 1
Thresholds for mapping loci underlying complex traits

	Suggestive linkage	Significant linkage
Mapping method	***P* value (lod)**	***P* value (lod)**
Lod score analysis in humans	1.7×10^{-3} (1.9)	4.9×10^{-5} (3.3)
Sib-pair method	7.4×10^{-4} (2.2)	2.2×10^{-5} (3.6)
QTL mapping		
Backcross	3.4×10^{-3} (1.9)	1.0×10^{-4} (3.3)
F2 intercross	3.4×10^{-3} (1.9)	1.0×10^{-4} (3.3)

hood has been proposed in the literature [15]. $L(H)$ is calculated for all the chromosomal intervals, and the lod score curve is illustrated based on the data of all the individuals by computer programs. The lod score curve tends to have protruding arcs between the plots of lod scores at marker loci (Fig. 3d). The peak position of the lod score curve gives the MLE of the map position θ of the QTL on the chromosome. The "one-log support interval" is obtained by drawing a horizontal line at 1 unit of lod score below the peak of the lod curve. The points of intersection of this line with the lod score curve are projected onto the genetic map and show the genomic interval at which the QTL should be found with a probability of 90 %.

1.4 Statistical Significance of the Linkage

When a quantitative phenotype appears to be associated with a particular marker allele in a simple association study, the statistical significance of the linkage is evaluated based on a significance level α (the probability of a false-positive observation by chance), for instance, $\alpha = 0.05$. However, if this type of association is tested for 200 genetic markers distributed on all the chromosomes, some of the markers will show the false-positive linkages by chance. Therefore, in genome-wide studies, a more stringent criterion is needed for the interpretation of linkage data. As genetic markers are in linkage groups, a proper criterion for the support of linkage cannot be determined simply. The *empirical significance level* is determined by computer simulation of the segregation of loci or by a permutation test of the phenotypic values among the individuals of the progeny. A program for the permutation test is implemented in some of the genetic analysis software packages, such as MapManager QTX [16]. Table 1 summarizes the significance levels for genome-wide studies on the complex traits proposed by Lander and Kruglyak [17]. They proposed that an association with a lod score of over 3.3 in experimental backcross studies should be termed a "significant linkage," and they recommended that a gene name not be given to any locus with a lod score below 3.3 in backcross studies. Some researchers feel that this standard may be too strict to detect minor alleles involved in multifactorial disease. Another proposal is that lod scores between 2.2 and 3.3 could be considered as "suggestive linkages."

2 Materials

2.1 Isolation of Genomic DNA from Mouse Tails (See Note 1)

1. Lysis buffer: 0.05 M Tris–HCl (pH 8.0), 0.1 M NaCl, 0.02 M EDTA, 1 % SDS. Lysis buffer was prepared by mixing appropriate volumes of stock solutions of 1 M Tris–HCl (pH 8.0), 5 M NaCl, 500 mM disodium EDTA (pH 8.0), and 10 % SDS (*see* **Note 2**).
2. Proteinase K: Lyophilized recombinant proteinase K was dissolved in double-distilled water at 10 mg/ml. Appropriate aliquots were frozen and stored at −20 °C.
3. Phenol solution saturated with 0.1 M Tris–HCl (pH 8.0): Commercial phenol for nucleic acid extraction was melted at 68 °C, 8-hydroxyquinoline (antioxidant) was added to a final concentration of 0.1 %, and the solution was extracted several times with an equal volume of 1.0 M Tris–HCl (pH 8.0) buffer. The liquefied phenol solution thus prepared was extracted with 0.1 M Tris (pH 8.0) buffer containing 0.2 % β-mercaptoethanol several times, until the pH of the aqueous phase was >7.6. The phenol solution can be stored at 4 °C under equilibration buffer for periods of up to 1 month (*see* **Note 3**).
4. Phenol:chloroform (1:1): The phenol solution as prepared above was mixed with an equal volume of chloroform (analytical grade).
5. Chloroform (analytical grade).
6. Ethanol (analytical grade).

2.2 Genotyping Microsatellite DNA Markers

1. Dye-labeled primer pairs for amplification of chromosomal microsatellite DNA markers: Pairs of 5′ fluorescent dye-labeled oligodeoxynucleotide and unlabeled oligodeoxynucleotide (Custom Primer Pair Part# 450056) were provided by Applied Biosystems (Foster City, CA). One of the four fluorescent dyes, FAM™ (5- or 6-carboxyfluorescein), NED™ (2′-chloro-5′-fluoro-7t, 8′ benzo-1, 4-dichloro-6-carboxy fluorescein), VIC™ (2′-chloro-7′phenyl-1, 4-dichloro-6-carboxy fluorescein), or PET™ (unpublished proprietary dye of Applied Biosystems), was selected for labeling (*see* **Note 4**).
2. Thermo-stable DNA polymerase (TaKaRa Taq™: Code No. R001AM, 5 units/μL) (*see* **Note 5**).

3. Deoxyribonucleoside-5′-triphosphates (dATP, dGTP, dCTP, dTTP): A mixture of four deoxyribonucleotide solutions, each at 2.5 mM adjusted to pH 7–9, was provided by the manufacturer of polymerase.
4. Optimized buffer for PCR amplification: Provided by the manufacturer of DNA polymerase (10× PCR buffer: 100 mM Tris–HCl pH 8.3, 500 mM KCl).
5. $MgCl_2$ solution (25 mM solution supplied by the manufacturer of polymerase).
6. Oligodeoxyribonucleotides labeled at 5′ with the fluorescence dye LIZ™ for size markers: GeneScan™ 500 LIZ® Size Standard (Cat# 4322682; Applied Biosystems, Foster City, CA; LIZ® is an unpublished proprietary dye of Applied Biosystems).
7. Formamide for DNA sequencing: Hi-Di™ Formamide (Cat# 4311320; Applied Biosystems, Foster City, CA).
8. GeneAmp 9700 thermal cycler (Applied Biosystems) or equivalent.
9. 3130 Genetic analyzer (Applied Biosystems) equipped with a 33 cm capillary array (Cat# 4319898; Applied Biosystems).
10. Windows-based work station equipped with GeneMapper software (Applied Biosystems).

2.3 Statistical Analysis

1. Windows-based personal computer installed with MAPMAKER/exe and MAPMAKER/qtl software [4, 5] (*see* **Note 6**).

3 Methods

3.1 Extraction of Genomic DNA from Mouse Tails

1. The terminal of the mouse tail was held using a forceps, and a piece of the tail (1.5 cm long) was cut off with surgical scissors. The small amount of bleeding stopped quickly, and it required no further treatment. Tail specimens may be stored in microtubes at −20 °C.
2. One milliliter of lysis buffer [0.05 M Tris–HCl (pH 8.0), 0.1 M NaCl, 0.02 M EDTA, 0.1 % SDS] was added to the tube containing a tail specimen, followed by 10 μL of proteinase K (10 mg/ml). After brief mixing, the tube was incubated at 55 °C for 16 h to digest the tail.
3. Tubes were centrifuged at top speed (17,000–20,000 × *g*) at room temperature for 10 min. The clear and viscous supernatant (500 μL) was transferred to a new tube.
4. To the supernatant, an equal volume of phenol solution was added. The tube was placed on a rotator or a horizontal shaker to aid in gentle mixing (1–2 rpm) for 10 min. Then, the tube

was centrifuged at top speed (17,000–20,000×*g*) for 3 min to obtain the clear upper aqueous phase. The supernatant (450 μL) was carefully transferred to a new tube. The supernatant was similarly extracted once with an equal volume of a mixture of phenol:chloroform (1:1) and once with chloroform.

5. Genomic DNA was precipitated by adding two volumes of ethanol, followed by centrifugation. The precipitate was washed once with 70 % ethanol and dried briefly in vacuo. The precipitate was rehydrated with 400 μL of water overnight. The concentration of DNA was determined by the absorbance at 260 nm. The concentration of DNA was adjusted to 5–10 μg/mL with 10 mM Tris–HCl (pH 7.5) containing 0.1 mM EDTA. The DNA solution was then ready for genotyping.

3.2 Analysis of the Polymorphism of Microsatellite DNA Markers

1. The PCR mixtures were added to the wells of a MicroAmp™ Optical 96-Well Reaction Plate (Part # N801-056; Applied Biosystems, Foster City, CA) by adding 3 μL of genomic DNA (8 μg/ml) and 3 μL of 2× reaction mixture. The final concentrations making up the reaction mixture were as follows: 107 nM each of the forward and reverse primers, 226 μM each of the four deoxynucleoside triphosphates (dATP, dGTP, dCTP, and dTTP), 8.1 mM Tris–HCl (pH 8.3), 40.5 mM KCl, 2.7 mM $MgCl_2$, 4 μg/mL DNA, and 0.032 units/μL *Taq* polymerase (*see* **Note** 7).
2. The PCR amplifications were carried out using a GeneAmp 9700 PCR System (Applied Biosystems). The thermal cycling consisted of an initial denaturation at 92 °C for 1 min, followed by 36 cycles of 92 °C for 1 min, 56–58 °C for 1.5 min, and 72 °C for 2 min, and a final extension at 72 °C for 10 min.
3. After the thermal cycles for amplification, 180 μL of double-distilled water was added to each well of the PCR plate. Subsequently, 2 μL of each diluted PCR mixture was transferred to the well of a 96-well PCR plate and mixed with 8 μL of LIZ®-labeled size marker DNA solution diluted in formamide (GeneScan™ 500 LIZ® Size Standard: Cat# 4322682; Applied Biosystems, Foster City, CA; LIZ® is an unpublished proprietary dye of Applied Biosystems). The mixture was heated at 96 °C for 5 min and then chilled at 4 °C for 5 min.
4. The PCR products were analyzed using an ABI Prism 3130xl DNA Analyzer (Applied Biosystems). The sizes of the amplified microsatellite DNA were determined using GeneMapper software (Applied Biosystems). A representative electrophoretic profile of a polymorphic microsatellite DNA fragment is shown in Fig. 4.

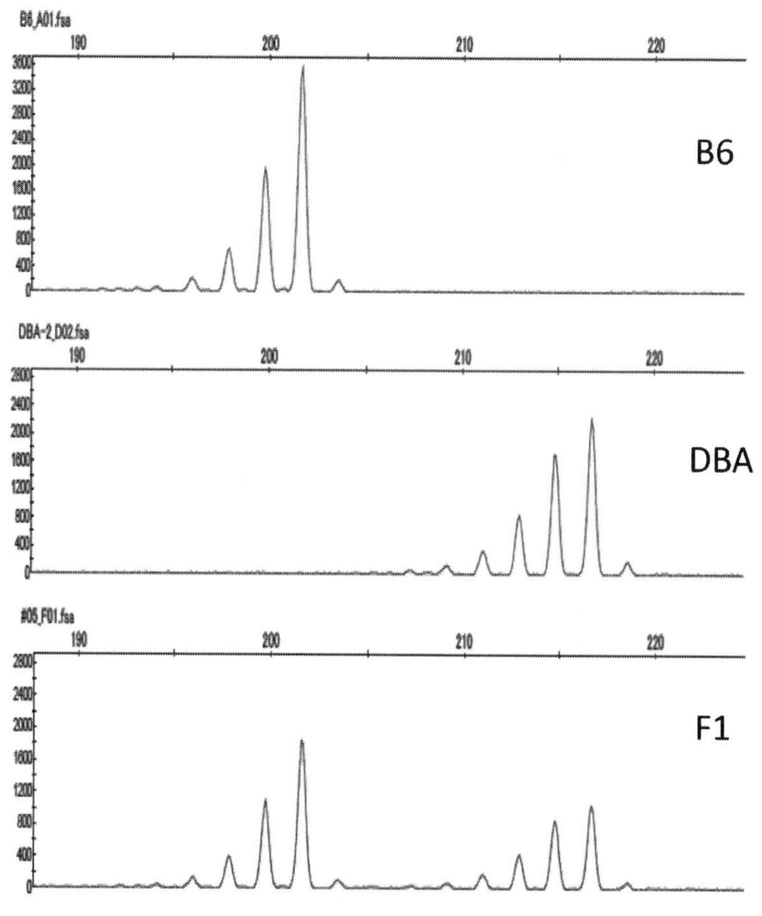

Fig. 4 Electrophoretic profiles of the polymorphic microsatellite DNA fragments (*D1Mit24*) from C57BL/6 (B6), DBA/2(DBA), and (C57BL/6 × DBA/2)F1 (F1) strains obtained by an ABI 3130xl genetic analyzer

3.3 Statistical Analysis: QTL Mapping

The most widely used tool for mapping quantitative traits in backcross studies has been the program developed by Lander et al. [4, 5]. The MAPMAKER program uses the maximum likelihood method. This method enables interval mapping for the potential QTL in the genomic interval between the markers. A data file needs to be written in a simple text using a one-byte alphanumeric character and symbols (ASCII code; *see* **Note 8**). The program is divided into two parts. One part is the MAPMAKER/EXP program for constructing genetic maps based on the observed segregation of the polymorphic genetic markers in the progeny of the experimental cross. The other part, the MAPMAKER/qtl program, is used for the genome-wide scanning of QTLs on the genetic map. Lod scores are calculated for putative QTLs to provide measures of support for the linkages. The details of the tutorial for using MAPMAKER are provided together with the sample data files. The following is an outline of the procedure of QTL mapping in the progeny of an experimental cross of mouse strains. Users of this program are advised to read the tutorials provided for the programs (*see* **Note 9**). The following guide to the procedure assumes that the researcher uses the MAPMAKER program on Windows-based personal computer.

1. *Preparation of data file for MAPMAKER*: An example of a data file for a progeny of backcross mice is shown in Fig. 5. The genotype/phenotype data for each individual mouse of the experimental backcross progeny need to be summarized in a simple text format. Researchers may have summarized the genotype/phenotype data using a standard spreadsheet program such as Microsoft EXCEL. In this case, data on the spreadsheet are easily converted to a simple text format.

   On a Microsoft EXCEL worksheet, the names of the genetic markers are listed in the cells of the first row. Because the loci of genetic markers have recently been precisely defined on the physical map of mouse chromosomes, the names of the genetic markers should be aligned in accordance with the physical map. The numbers of individuals in the backcross progeny are listed on the cells of the first column. Then, the genotypes of each individual are listed in the corresponding cells of the following rows. Affection status or quantitative phenotypes of each individual are listed in the last column of the spreadsheet. Names of the genetic markers and quantitative phenotypes should be abbreviated using eight letters or less, and each name should begin with an asterisk (*). The names of the genetic markers and phenotypes should start with any character but not with numbers. Note that genotypes with respect to any genetic markers in the progeny (either homozygous or heterozygous) should be represented by common symbols using single capital letters. Missing data should be represented with a hyphen (-). To convert the data sheet into the text format for MAPMAKER, all the cells in the spreadsheet are selected and copied and pasted onto a new spreadsheet while exchanging the rows and columns by selecting the "Paste Special" command from the menu. Subsequently, the spreadsheet is exported in a simple text format by selecting the "tab-spaced text format" option. The exported text is then edited by using a simple text editor to conform the data format for MAPMAKER, as shown in Fig. 5. The first line declares the data type of a mating (F2 intercross); the second line indicates the number of progeny (200 mice), the total number of loci analyzed (18), and the total number of quantitative loci to be analyzed (in this case, one), followed by the definition of symbols for the genotypes (either homozygous or heterozygous)

spreadsheet format

Mouse No	D1Mit66	D1Mit20	D1Mit24	D1Mit10	D1Mit26	D1Mit102	D3Mit256	D3Mit19	PheVal
1	B	B	B	B	B	H	A	A	0.38
2	H	H	H	H	H	H	H	H	0.55
3	H	H	B	B	B	B	B	B	-0.32
4	H	H	H	H	H	H	A	B	-0.2
5	H	A	H	H	H	H	B	H	0.6
6	H	A	A	A	A	A	H	H	0.22
7	H	H	H	H	H	H	A	A	0.31
8	H	H	B	B	B	B	B	B	0.44

MAPMAKER/qtl format

Data type F2 intercross

200 18 1 A=A B=B H=H - = -

*D1Mit66	B	H	H	H	H	H	H	H
*D1Mit20	B	H	H	H	A	A	H	H
*D1Mit24	B	H	B	H	H	A	H	B
*D1Mit10	B	H	B	H	H	A	H	B
*D1Mit26	B	H	B	H	H	A	H	B
*D1Mit102	H	H	B	H	H	A	H	B
*D3Mit256	A	H	B	A	B	H	A	B
*D3Mit19	A	H	B	B	H	H	A	B
*PneVa	0.38	0.55	-0.32	-0.2	0.6	0.22	0.31	0.44

Fig. 5 An example of a data format to be analyzed by MAPMAKER/EXE and MAPMAKER/QTL programs. Data in a spreadsheet format containing genotypes and phenotypes of each individual mouse are converted into a text that conforms to the MAPMAKER format. In the MAPMAKER format, the *first line* states the type of mating; the *second line* indicates the number of progeny (200 mice), the total number of marker loci analyzed (18), and the total number of quantitative loci (1) to be analyzed. The names of each locus should be preceded by an asterisk (*)

as well as for the missing data. In the subsequent lines, the genotype and phenotype data of the individuals are aligned sequentially. Names of the genetic markers or the names of the phenotypes are coded at the beginning of each line, and each name must be marked with an asterisk (*). The data file is then copied into the "MAPMAKER" directory. The name of the data file thus prepared by using a text editor may have an extension such as ".txt." In this case, the file name is renamed

An example of MAPMAKER commands in a batch file (Testdata.pre)

```
unit cm
cent func kosambi
photo testdata.out
print names on
triple error detection on
informativeness criteria 1.0 100
make chromosome chrom1 chrom3
sequence D1Mit24 D1Mit149
anchor chrom1
sequence D3Mit227 D3Mit256
anchor chrom3
sequence all
two point
save
```

Fig. 6 An example of a series of MAPMAKER/EXP commands for constructing a genetic map (a framework for QTL mapping) on mouse chromosome 1 and 3. A batch file with these commands is prepared in a simple text format to automate the process

on the command line to have the ".raw" extension: for example, "testdata.raw."

2. *Batch file for MAPMAKER/EXP*: Before conducting QTL mapping, it is necessary to generate a genetic map with respect to the genetic markers on each chromosome, which is subsequently used as a framework for QTL mapping on the chromosome. This process is conducted by the MAPMAKER/EXP program. The data files used in MAPMAKER/qtl are also produced from the original raw data file in this process. The process consists of three steps:
   (a) Assignment of each genetic marker into linkage groups.
   (b) Definition of chromosomes.
   (c) Constructing genetic maps on chromosomes based on the experimental data.

   To conduct these steps smoothly, a "batch file" should be prepared to automate the necessary steps. An example of a batch file for processing the raw file shown in Fig. 6 is written in a simple text format and is saved with the same name as used for a raw data file with the extension ".pre" (testdata.pre).
3. *Constructing genetic linkage map with MAPMAKER/EXE*: Two text files, testdata.raw and testdata.pre, are copied and pasted into the MAPMAKER directory in which the linkage analysis is to be conducted. Then, the following procedures are conducted on command lines. Knowledge of the basic commands

dealing with files and directories is required. From the Windows menu, "Command Prompt" is selected to start the command lines. The following is an example of the inputs and outputs using a Windows XP-based PC (Vostro 1720; DELL Inc.) (*see* **Note 10**). Throughout this example, all inputs to the computer are presented using ***bold italics***, while MAPMAKER outputs are shown in `ordinary type`.

```
C:¥Documents and Settings¥UserName> cd ..

C:¥Documents and Settings> cd ..

C:¥ cd MAPMAKER

C:¥MAPMAKER>dir testdata.*

2013/07/03 17:13        270 testdata.pre

2013/07/03 17:00      9,862 testdata.txt
```

The extensions of the file name are changed to the appropriate ones.

```
C:¥MAPMAKER> ren testdata.txt testdata.raw
```

Now that the files are ready, it is time to start the procedure. MAPMAKER/EXP is started by typing "mapmaker" at the command prompt.

```
C:¥MAPMAKER>mapmaker
*********************************************************************
* Output from:                                Sat Aug 25 00:02:38 2013
*
*                          MAPMAKER/EXP
*                         (version 3.0b)
*
*********************************************************************
Type 'help' for help.
Type 'about' for license, non-warranty, and support information.
1> prepare data testdata.raw
data from 'TESTDATA.RAW' are loaded
  F2 intercross data  (200 individuals, 18 loci)
'photo' is on: file is 'TESTDATA.OUT'
5> print names on
'print names' is on.
6> triple error detection on
'triple error detection' is on.
7> informativeness criteria 1.0 100
Informativeness Criteria: min Distance 1.0, min #Individuals 100
8> make chromosome chrom1 chrom3
chromosomes defined: chrom1 chrom3
9> sequence D1Mit24 D1Mit149
sequence #1= D1Mit24 D1Mit149
10> anchor chrom1
D1Mit24  - anchor locus on chrom1
D1Mit149 - anchor locus on chrom1
chromosome chrom1 anchor(s): D1Mit24 D1Mit149
1   sequence D3Mit227 D3Mit256
sequence #2= D3Mit227 D3Mit256
12> anchor chrom3
D3Mit227 - anchor locus on chrom3
D3Mit256 - anchor locus on chrom3
chromosome chrom3 anchor(s): D3Mit227 D3Mit256
13> sequence all
sequence #3= all
14> two point
two-point data are available.
15> save
```

```
saving genotype data in file 'TESTDATA.DAT'... ok
saving map data in file 'TESTDATA.MAP'... ok
saving two-point data in file 'TESTDATA.2PT'... ok
saving traits data in file 'TESTDATA.TRA'... ok

16>
        ...end of input file...
```

The above outputs are the results of an automated process programmed in a batch file "*testdata.pre*." The following are the manual inputs on the command lines and the outputs. All the genetic markers are selected by "sequence" command, and each genetic marker is assigned to each chromosome.

```
17> sequence all
sequence all
sequence #4= all

17> assign
D1Mit24  - anchor locus on chrom1...cannot re-assign
D1Mit149 - anchor locus on chrom1...cannot re-assign
D3Mit227 - anchor locus on chrom3...cannot re-assign
D3Mit256 - anchor locus on chrom3...cannot re-assign
D1Mit66  - assigned to chrom1 at LOD 13.6
D1Mit20  - assigned to chrom1 at LOD 49.7
```

.... lines of outputs follow.....

```
D3Mit10  - assigned to chrom3 at LOD 52.9
D3Mit19  - assigned to chrom3 at LOD 24.1
```

To construct a genetic map on each chromosome, the most plausible order of genetic markers is estimated. First, the genetic markers assigned to chromosome 1 are selected and subjected to three-point analysis.

```
18> sequence chrom1
sequence #5= chrom1: assigned


19> three point
Linkage Groups at min LOD 3.00, max Distance 37.2
Triplet criteria: LOD 3.00, Max-Dist 37.2, #Linkages 2
'triple error detection' is on.
counting...84 linked triplets in 1 linkage group

                                   log-likelihood differences
 count  markers                        a-b-c  b-a-c  a-c-b
    1:  D1Mit66 D1Mit20 D1Mit24          0.00 -10.23 -34.00
    2:  D1Mit66 D1Mit20 D1Mit10          0.00  -2.30 -44.88
    3:  D1Mit66 D1Mit24 D1Mit10          0.00 -28.97  -8.84
```

.... lines of outputs follow.....

```
   83:  D1Mit14 D1Mit166 D1Mit155        0.00 -35.46  -6.74
   84:  D1Mit149 D1Mit166 D1Mit155       0.00 -16.60 -11.23
```

The most plausible order of markers on chromosome 1 is estimated by an "order" command.

```
20> order
Linkage Groups at min LOD 3.00, max Distance 37.2
Starting Orders: Size 5, Log-Likelihood 3.00, Searching up to 50 subsets
Informativeness: min #Individuals 100, min Distance 1.0
Placement Threshold-1 3.00, Threshold-2 2.00, Npt-Window 7
======================================================================
======
Linkage group 1, 11 Markers:
   1 D1Mit66       2 D1Mit20       3 D1Mit24       4 D1Mit10       5 D1Mit26
   6 D1Mit102       7 D1Mit159       8 D1Mit14       9 D1Mit149       10
D1Mit166
  11 D1Mit155

All markers are informative...
Searching for a starting order containing 5 of all 11 loci...
Got one at log-likelihood 999716.38

Placing at log-likelihood threshold 3.00...
Start:   3 4 6 8 9
3pt:     3 4 (5) 6 8 9
```

.... lines of outputs follow.....

```
    8  D1Mit14      7.9 cM
    9  D1Mit149     6.5 cM
   10  D1Mit166     5.1 cM
   11  D1Mit155   ----------
                  78.3 cM   11 markers   log-likelihood= -478.87

order1= D1Mit66 D1Mit20 D1Mit24 D1Mit10 D1Mit26 D1Mit102 D1Mit159 D1Mit14
D1Mit149 D1Mit166 D1Mit155
other1=
======================================================================
======
```

The most plausible order of genetic markers is selected, and the corresponding genetic map is set as the framework for QTL mapping.

```
21> sequence order1
sequence #6= chrom1: order1

22> framework chrom1
setting framework for chromosome chrom1...
==============================================================================
chrom1 framework:

 Markers          Distance
   1  D1Mit66       7.9 cM
   2  D1Mit20      17.9 cM
```

.... lines of outputs follow.....

```
  10  D1Mit166      5.1 cM
  11  D1Mit155    ----------
                   78.3 cM   11 markers   log-likelihood= -478.87
==============================================================================
```

The same procedure is applied to the markers on chromosome 3

```
23> sequence chrom3
sequence #7= chrom3: assigned

24> three point
Linkage Groups at min LOD 3.00, max Distance 37.2
Triplet criteria: LOD 3.00, Max-Dist 37.2, #Linkages 2
'triple error detection' is on.
counting...22 linked triplets in 1 linkage group

                                         log-likelihood differences
 count  markers                              a-b-c  b-a-c  a-c-b
    1:  D3Mit203 D3Mit227 D3Mit22             0.00 -21.57 -16.27
    2:  D3Mit203 D3Mit227 D3Mit9              0.00 -16.32 -23.38
```

.... lines of outputs follow.....

```
   21:  D3Mit9 D3Mit256 D3Mit19               0.00 -18.58 -11.71
   22:  D3Mit10 D3Mit256 D3Mit19              0.00 -16.21 -20.58
```

```
25> order
Linkage Groups at min LOD 3.00, max Distance 37.2
Starting Orders: Size 5, Log-Likelihood 3.00, Searching up to 50 subsets
Informativeness: min #Individuals 100, min Distance 1.0
Placement Threshold-1 3.00, Threshold-2 2.00, Npt-Window 7
=======================================================================
======
Linkage group 1, 7 Markers:
  12 D3Mit203    13 D3Mit227    14 D3Mit22    15 D3Mit9    16 D3Mit10
  17 D3Mit256    18 D3Mit19

All markers are informative...
Searching for a starting order containing 5 of all 7 loci...
Got one at log-likelihood 6.31

Placing at log-likelihood threshold 3.00...
Start:  12 13 14 16 18
3pt:    12 13 14 16 (17) 18
```

.... lines of outputs follow.....

```
   16  D3Mit10     15.6 cM
   17  D3Mit256    18.6 cM
   18  D3Mit19    ----------
                  65.2 cM   7 markers   log-likelihood= -385.14

order1= D3Mit203 D3Mit227 D3Mit22 D3Mit9 D3Mit10 D3Mit256 D3Mit19
other1=
=======================================================================
======

26> sequence order1
sequence #8= chrom3: order1

27> framework chrom3
setting framework for chromosome chrom3...
=======================================================================
======
```

```
chrom3 framework:

  Markers          Distance
   12  D3Mit203     10.7 cM
   13  D3Mit227      9.3 cM
   14  D3Mit22       4.4 cM
   15  D3Mit9        6.8 cM
   16  D3Mit10      15.6 cM
   17  D3Mit256     18.6 cM
   18  D3Mit19     ----------
                   65.2 cM   7 markers   log-likelihood= -385.14
======================================================================
======
```

The process using MAPMAKER/EXP ends here. The program is terminated, and the calculated data are saved in files automatically.

```
28> quit
save data before quitting? [yes] yes
saving map data in file 'TESTDATA.MAP'... ok
saving three-point data in file 'TESTDATA.3PT'... ok

      ...goodbye...

C:¥MAPMAKER>
```

4 *QTL mapping using MAPMAKER/qtl*: Subsequently, MAPMAKER/qtl is activated by typing “qtl” on the command line.

```
C:¥MAPMAKER>qtl

**********************************************************************
* Output from:                                  Sat Aug 25 00:07:02 2013
*
*                            MAPMAKER/QTL
*                           (version 1.1b)
*
**********************************************************************
Type 'help' for help.
Type 'about' for license, non-warranty, and support information.
1> load testdata
data files 'testdata' and 'testdata.traits' are loaded.
(200 intercross progeny, 19 loci, 1 trait)
Unable to load any saved QTL data.
The process of inputs and outputs are recorded in a test file by "photo"
command.
1> photo testdata.out
'photo' is on: file is 'TESTDATA.OUT'
Set an option to use the name of the genetic markers in the outputs.
2> print names on
'print names' is on.

3> sequence [all]
The sequence is now '[all]'
```

The algorithms on which MAPMAKER/QTL is programmed assume that values of the quantitative trait vary across the population following a normal distribution. Thus, the first step is to examine the trait data and determine how well they fit this assumption. If they do not closely follow a normal distribution, it is advised that the data be transformed into a "derived trait" which more closely fits the assumption.

```
4> trait 1
The current trait is now: 1 (PneVa)


5> show trait


Trait 1 (PneVa):

----------------------------------------------------------------------
------
distribution:                               quartile |    fraction within n
deviations:
mean   sigma   skewness  kurtosis  ratio    |  1/4    1/2   1     2     3
0.31   0.33    -0.04     0.03      0.91     |  0.18   0.40  0.71  0.95
1.00
----------------------------------------------------------------------
------

 -0.34  |****
 -0.18  |**************
 -0.02  |*********************
  0.15  |*********************************************
  0.31  |******************************************************
  0.47  |************************************************************
  0.63  |*****************************************
  0.80  |*********************
  0.96  |************
  1.12  |*****
```

Distribution of the phenotype data is demonstrated by a histogram plot. For a more detailed tutorial, *see* the instruction manual for MAPMAKER/qtl. Then, the selected chromosomes are scanned for the presence of the putative QTL.

```
9> scan

QTL maps for trait 1 (PneVa):
Sequence: [all]
LOD threshold: 2.00  Scale: 0.25 per '*'
No fixed-QTLs.
Scanned QTL genetics are free.

POS     WEIGHT  DOM      %VAR  LOG-LIKE |
--------------------------------------| D1Mit66-D1Mit20 8.5 cM
0.0     -0.038  -0.012   0.7%    0.317  |
2.0     -0.035  -0.015   0.7%    0.276  |
```

.... lines of outputs follow.....

```
10.0    -0.088   0.044   4.3%    1.764  |
--------------------------------------| D3Mit227-D3Mit22 10.1 cM
0.0     -0.091   0.044   4.5%    1.980  |
2.0     -0.103   0.055   5.8%    2.307  | **
4.0     -0.109   0.059   6.5%    2.540  | ***
6.0     -0.110   0.059   6.7%    2.675  | ***
8.0     -0.108   0.056   6.5%    2.722  | ***
10.0    -0.103   0.052   6.0%    2.694  | ***
--------------------------------------| D3Mit22-D3Mit9 4.5 cM
0.0     -0.103   0.052   6.0%    2.691  | ***
2.0     -0.106   0.058   6.4%    2.843  | ****
4.0     -0.107   0.063   6.6%    2.950  | ****
--------------------------------------| D3Mit9-D3Mit10 7.2 cM
0.0     -0.107   0.065   6.6%    2.971  | ****
2.0     -0.105   0.066   6.5%    2.769  | ****
4.0     -0.100   0.066   6.0%    2.480  | **
6.0     -0.091   0.062   5.0%    2.086  | *
--------------------------------------| D3Mit10-D3Mit256 18.1 cM
0.0     -0.082   0.054   4.1%    1.795  |
```

.... lines of outputs follow.....

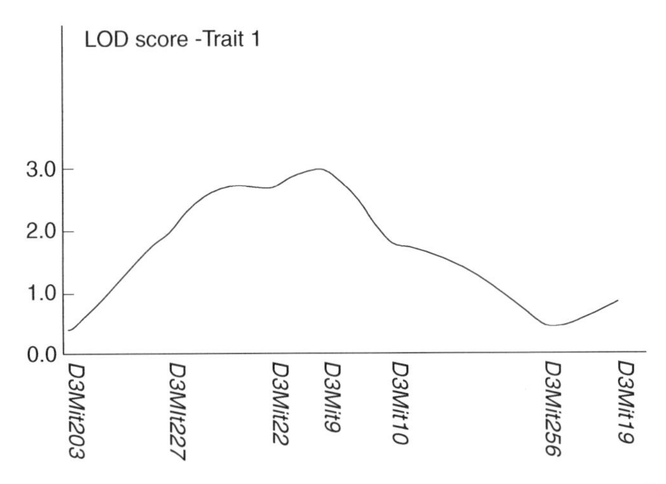

Fig. 7 An example of a lod score curve illustrated by the MAPMAKER/QTL program

```
18.0   -0.046   0.131   5.3%    1.890  |
20.0   -0.044   0.130   5.1%    2.025  | *
-------------------------------------|

Results have been stored as scan number 1.
```

The lod curve is generated in PostScript format by the following command. The resulting PostScript file can be viewed and manipulated by using a graphic software package such as Adobe Illustrator. An example of the lod curve obtained by this process is shown in Fig. 7.

```
10> draw scan

scan 1.1 saved in PostScript file 'scan1_1.ps'
```

The parameters for the peak position are obtained by the following command.

```
11> show peaks

LOD score peaks for scan 1.1 of trait 1 (PneVa).
Sequence: [all]
No fixed-QTLs.
Scanned QTL genetics are free.
Peak Threshold: 2.00  Falloff: -2.00

======================================================================
======
QTL-Map for peak 1:
Confidence Interval:  Left Boundary= D3Mit203-D3Mit227 + 6.0
                  Right Boundary= D3Mit10-D3Mit256 + 12.0

INTERVAL             LENGTH  QTL-POS  GENETICS    WEIGHT  DOMINANCE
D3Mit9-D3Mit10        7.2     0.0      free      -0.1065  0.0645

chi^2= 13.681  (2  D.F.)       log-likelihood= 2.97
mean= 0.386   sigma^2= 0.099   variance-explained= 6.6 %
======================================================================
======
QTL-Map for peak 2:
Confidence Interval:  Left Boundary= D3Mit203-D3Mit227 (off end)
                  Right Boundary= D3Mit256-D3Mit19 (off end)

INTERVAL             LENGTH  QTL-POS  GENETICS    WEIGHT  DOMINANCE
D3Mit256-D3Mit19      21.9    20.0     free      -0.0443  0.1295

chi^2= 9.327  (2  D.F.)       log-likelihood= 2.03
mean= 0.295   sigma^2= 0.100   variance-explained= 5.1 %
======================================================================
======
```

The process using MAPMAKER/qtl ends here, and the output files are automatically generated.

```
12> quit
save data before quitting? [yes] yes
Now saving TESTDATA.QTLS...
Now saving TESTDATA.TRAITS...

            ...goodbye...
```

4 Notes

1. Researchers may prefer to use commercial kits for genomic DNA isolation. The advantage of applying this old traditional protocol is that the relatively large amount of purified DNA that is required for automated liquid handling can be obtained. The effect of polysaccharide contaminations that may inhibit DNA polymerase is minimized at the working concentration of genomic DNA (5–10 μg/ml) in the PCR mixture.
2. For 1 M Tris stock solution, 121.1 g of Tris base (2-amino-2-hydroxymethyl-propane-1,3-diol of molecular biology grade) is dissolved in 800 ml of distilled water, and an appropriate volume (approximately 42 ml) of concentrated HCl is added to adjust the pH (8.0), bringing the final volume of the solution to 1,000 ml. The Tris stock solution is dispensed in aliquots and sterilized by autoclaving. For 5 M NaCl stock solution, 292.2 g of NaCl is dissolved in 800 ml of distilled water, bringing the final volume to 1,000 ml, and sterilized by autoclaving. For 500 mM EDTA stock solution, 186.1 g of disodium ethylenediaminetetraacetate $2H_2O$ is dissolved in 800 ml of distilled water, and the pH is adjusted to 8.0 by adding an appropriate amount of sodium hydroxide pellets (approximately 20 g), bringing the final volume to 1,000 ml. The stock solution is dispensed in aliquots and sterilized by autoclaving. For 10 % SDS stock solution, 100 g of electrophoresis-grade SDS is dissolved in 900 ml of distilled water and heated to 68 °C to assist complete dissolution, and the pH is adjusted to 7.2 by adding a few drops of concentrated HCl, bringing the total volume to 1,000 ml. The stock solution is dispensed into aliquots. There is no need to sterilize SDS solution.
3. Phenol is highly toxic and corrosive. Details of the precautions for handling phenol reagent are described in the literature [18].
4. Fluorescence dyes detected by the genetic analyzer differ by the capillary electrophoresis machines for genetic analysis. A series of genetic analyzers produced by Applied Biosystems Inc. (Foster City, CA) detects the fluorescence dyes FAM™, VIC™, NED™, and PET™. Applied Biosystems accepts requests for custom synthesis of PCR primer pairs, one primer of which is labeled by one of these four dyes via their website (https://products.appliedbiosystems.com/ab/en/US/adirect/ab?Cmd=ABLogin).
5. There is no particular preference for the type of thermostable DNA polymerase. The buffer solution optimized for PCR and mixed deoxynucleoside triphosphate (dNTP) solution

are usually supplied by the manufacturer. It is preferable that the concentration of magnesium chloride be optimized for each primer pair.

6. The MAPMAKER/EXP and MAPMAKER/qtl programs are available from the FTP site of the Broad Institute (http://www.broadinstitute.org/ftp/distribution/software/). MAPMAKER can be installed in Windows-based PCs running MS-DOS, Sun-SPARCstations, A/UX Macintoshes, and Linux-based computers. A detailed guide to installing MAPMAKER 3.0 and MAPMAKER/qtl 1.1 on a Windows-based PC is also available from this site.
7. As the volume of reaction mixture (6 μL) was small, PCR was sensitive to the change of salt concentrations in the reaction mixture caused by drying while setting up multiple reactions. The final salt concentrations shown here were lower (81 %) than those recommended by the manufacturer of polymerase. In case of frequent failure of PCR, a larger volume of reaction mixture is recommended.
8. In the Windows-based PC, a simple text file is written using the application "Note Pad" or a "Simple Text" program and saved by using UNICODE (UTF-8). Data files exported from the Microsoft EXCEL program are edited by the Simple Text program to conform to the format for MAPMAKER and need to be checked by the "type (file name)" command at the command prompt.
9. Tutorials for using MAPMAKER/EXP and MAPMAKER/qtl are provided by Dr. E. Lander (The Lander Lab, Attn: MAPMAKER Support, Whitehead Institute for Biomedical Research, Cambridge Center, Cambridge, MA 02142, USA). PDF files are also distributed on the Internet.
10. MAPMAKER programs tend to be unstable when using the command prompt in the recent versions of the Windows operating system. Terminating background activities such as those conducted by security programs may help MAPMAKER to run more stably. This can be done by booting Windows in "diagnostic mode."

Acknowledgements

This work was supported by a Grant in Aid for Scientific Research C from the Ministry of Education, Science, Technology, Sports and Culture, Japan.

References

1. Falconer DS , Mackay TFC (1996) Threshold character. In: Introduction to quantitative genetics, Part 18 4th edn, Pearson Education Limited, Essex England, pp 295–311
2. Stansfield WD (1982) Quantitative genetics and breeding principles. In: Theory and problems of genetics 4th Edn, Schaums Outline Series. McGraw-Hill, New York, NY, pp 213–247
3. Stansfield WD (1982) Theory and problems of genetics. McGraw-Hill, New York, NY
4. Lander ES, Green P, Abrahamson J, Barlow A, Daly MJ, Lincoln SE, Newburg L (1987) MAPMAKER: an interactive computer package for constructing and natural populations. Genomics 1:174–181
5. Peterson A, Lander ES, Hewitt J, Peterson S, Tanksley S (1988) Resolution of quantitative traits into Mendelian factors using a complete RFLP linkage maps. Nature 335: 721–726
6. Kono DH, Burlingame RW, Owens DG, Kuramochi A, Balderas RS, Balomenos D, Theofilopoulos AN (1994) Lupus susceptibility loci in New Zealand mice. Proc Natl Acad Sci U S A 91:10168–10172
7. Drake CG, Babcock SK, Palmer E, Kotzin BL (1994) Genetic analysis of the NZB contribution to lupus-like autoimmune disease in (NZB x NZW) F1 mice. Proc Natl Acad Sci U S A 91:4062–4066
8. Rozzo SJ, Allard JD, Choubey D, Vyse TJ, Izui S, Peltz G, Kotzin BL (2001) Evidence for an interferon-inducible gene, *Ifi202*, in the susceptibility to systemic lupus. Immunity 15: 435–443
9. Hirose S, Sekigawa I, Ozaki S, Sato H, Shirai T (1984) Genetic regulation of hypergammaglobulinaemia and the correlation to autoimmune traits in (NZB x NZW) F1 hybrid. Clin Exp Immunol 58:694–702
10. Ochiai K, Ozaki S, Tanino A, Watanabe S, Ueno T, Mitsui K, Toei J, Inada Y, Hirose S, Shirai T, Nishimura H (2000) Genetic regulation of anti-erythrocyte autoantibodies and splenomegaly in autoimmune hemolytic anemia-prone New Zealand Black mice. Int Immunol 12:1–8
11. Jiang Y, Hirose S, Hammano Y, Kodera S, Tsurui H, Abe M, Terashima K, Ishikawa S, Shirai T (1997) Mapping of a gene for the increased susceptibility of B1 cells to Mott cell formation in murine autoimmune disease. J Immunol 158:992–997
12. Fujii T, Iida Y, Yomogida M, Ikeda K, Haga T, Jikumaru Y, Ninami M, Nishimura N, Kodera Y, Inada Y, Shirai T, Hirose S, Nishimura H (2006) Genetic control of the spontaneous activation of CD4+ Th cells in systemic lupus erythematosus-prone (NZB x NZW) F1 mice. Genes Immun 7(8):647–654
13. Fujii T, Hou R, Sato-Hayashizaki A, Obata M, Ohtsuji M, Ikeda K, Mitsui K, Kodera Y, Shirai T, Hirose S, Nishimura H (2011) Susceptibility loci for the defective foreign protein-induced tolerance in New Zealand Black mice: implication of epistatic effects of Fcgr2b and Slam family genes. Eur J Immunol 41(8):2333–2340.
14. Ott J (1999) Complex traits. In: Analysis of human genetic linkage, 3rd Edn. The Johns Hopkins University Press, Baltimore MD, pp 306–329
15. Lander ES, Botstein D (1989) Mapping Mendelian factors underlying quantitative traits using RFLP linkage maps. Genetics 121:185–199
16. Manly KF, Cudmore RH Jr, Meer JM (2001) Map Manager QTX, cross-platform software for genetic mapping. Mamm Genome 12(12):930–932
17. Lander ES, Kruglyak L (1995) Genetic dissection of complex traits: guidelines for interpreting and reporting linkage results. Nat Genet 11:241–247
18. Sambrook J, Fritsch EF, Maniatis T (1989) Isolation of high-molecular-weight DNA from mammalian cells. In: Ford N, Nora C, Ferguson M (eds) Molecular cloning: a laboratory manual, vol 1. Cold Spring Harbor Laboratory Press, Cold Spring Harbor, NY, p 914

Chapter 14

Bayesian Systems-Based Genetic Association Analysis with Effect Strength Estimation and Omic Wide Interpretation: A Case Study in Rheumatoid Arthritis

Gábor Hullám, András Gézsi, András Millinghoffer, Péter Sárközy, Bence Bolgár, Sanjeev K. Srivastava, Zsuzsanna Pál, Edit I. Buzás, and Péter Antal

Abstract

Rich dependency structures are often formed in genetic association studies between the phenotypic, clinical, and environmental descriptors. These descriptors may not be standardized, and may encompass various disease definitions and clinical endpoints which are only weakly influenced by various (e.g., genetic) factors. Such loosely defined complex intermediate clinical phenotypes are typically used in follow-up candidate gene association studies, e.g., after genome-wide analysis, to deepen the understanding of the associations and to estimate effect strength.

This chapter discusses a solid methodology, which is useful in such a scenario, by using probabilistic graphical models, namely, Bayesian networks in the Bayesian statistical framework. This method offers systematically scalable, comprehensive hierarchical hypotheses about multivariate relevance. We discuss its workflow: from data engineering to semantic publication of the results. We overview the construction, visualization, and interpretation of complex hypotheses related to the structural analysis of relevance. Furthermore, we illustrate the use of a dependency model-based relevance measure, which takes into account the structural properties of the model, for quantifying the effect strength. Finally, we discuss the "interpretational" or translational challenge of a genetic association study, with a focus on the fusion of heterogeneous omic knowledge to reintegrate the results into a genome-wide context.

Key words Genetic association studies, Detailed phenotyping, Bayesian networks, Bayesian multilevel analysis of variance, Bayesian structure-based effect strength estimation, Gene prioritization

1 Introduction

Rapid development of omic measurement techniques, particularly the spread of high-throughput genotyping and sequencing technologies, led to extreme diversity in the way studies define phenotypes: from the dichotomous case–control disease descriptor to tissue specific gene, proteomic, metabolomic expression profiles, which are also present as phenotypes [1]. There are many motivating factors to

Shunichi Shiozawa (ed.), *Arthritis Research: Methods and Protocols*, Methods in Molecular Biology, vol. 1142, DOI 10.1007/978-1-4939-0404-4_14, © Springer Science+Business Media New York 2014

bridge this gap, such as the modest success of case–control genome-wide association studies, clinical considerations of under- or ill-defined diseases, growing number of cohort studies with detailed epidemiologic, lifestyle and environmental data, and increasing computational power coupled with better algorithms [2, 3]. Currently, such detailed phenotyping is typically used to clarify the associations found in follow-up candidate gene association studies (CGAS), and partial genome screening studies (PGSS) performed after genome-wide analysis (GWA). However, the analysis of the relevance of predictors with respect to this set of phenotypic, clinical, and environmental descriptors is a challenging task, as these descriptors themselves are typically strongly interdependent, while the effects of the predictors are frequently found to be relatively weak and contextual.

A candidate for this task, the probabilistic graphical model (PGM) class provides a unifying framework for systems-based modeling and data analysis in computational biomedicine [4]. Specifically, this framework offers a principled foundation for multivariate association and interaction analysis, which is related to the feature subset selection problem in machine learning and computational statistics [5]. PGMs and its subclass: Bayesian networks (BNs), were quickly adopted in genetics and in genetic association studies (GASs), partly because of their causal aspects, and partly because of their systems-based foundations [6–10].

PGMs were also applied in genome-wide association studies (GWASs) to cope with interactions [10–13]. Although PGMs provide an attractive theoretical foundation for the systems based analysis of a network of phenotypic, environmental, and heterogeneous omic (e.g., genetic) variables, the complexity of the PGMs is much higher than that of the complexity of other multivariate GAS methods. In other words, the computational requirements of the general application of PGMs and the necessary sample size (in the frequentist statistical framework) are typically higher. An additional challenge is the explicit modeling of the interdependency structure of the target (or outcome) variables. These targets are typically clinical variables or disease state descriptors, which frequently show much stronger statistical dependencies than that of the genetic predictors. We proposed a PGM-based association analysis methodology called Bayesian multilevel analysis (BMLA) in the Bayesian statistical framework to cope with this challenge (for an overview, *see* ref. 14; for applications in GASs, *see* ref. 15–23). The Bayesian statistical framework is gaining popularity in biomarker discovery as well as in GASs [24, 25] because of the following factors. It is based primarily on model averaging instead of model selection in the frequentist statistical framework. The advantage of model averaging is that it provides an inherent, normative solution for multiple hypothesis testing [24]. It allows the principled incorporation of a wide range of informative prior knowledge into the statistical inference, which can be essential to preserve power [26–28]. The consequences of the statistical decisions, i.e., losses, gains, and costs, can be systematically combined with the results of the Bayesian statistical inference in the Bayesian decision theoretic framework, yielding optimal decisions [29–31].

The relatively high model complexity of PGMs and BNs poses a computational challenge to perform full-fledged Bayesian inference, although special methods relevant for Bayesian multivariate association analysis are available [7, 16, 32, 33]. Further challenging problems are the selection of noninformative prior distributions [34], the estimation of effect strength [35], and the use of omic level similarity and network knowledge in the interpretation and translation of the results [36]. In this chapter, we discuss these issues and provide a guide to the workflow of applying BNs in the Bayesian statistical framework in a CGAS context with complex, multiple phenotypic variables. We focus on the following steps in the workflow.

1. Prior distribution selection for PGM model structures and parameters.
2. Visualization of the results in Bayesian multilevel analysis using conditional relevance, relevance maps, relevance trees, and interaction maps.
3. Effect strength estimation using Bayesian, Markov blanket conditioned odds ratios.
4. Disease specific gene prioritization of CGAS/PGSS results.

2 Design Considerations for Bayesian Network-Based CGAS

The full-fledged Bayesian application of PGMs to explore the relevance of biomarkers, including genetic variants, with respect to detailed phenotypic descriptors encompasses both the structural exploration of the joint dependency model of all the variables and their effect strength estimations. This focused application is typical in hypothesis-driven CGASs, which means that both the selection of phenotypic (outcome) variables and genetic variables (predictors) is an open issue together with selection of an appropriate sample size. The general principles of CGAS study design, such as those collected below, can be applied despite the statistical and computational specialties of PGMs potentially influencing these decisions, because the application of PGMs is currently not exclusive and it corresponds more to the exploratory data analysis, which is typically checked by standard frequentist methods [34]. General CGAS study design guides cover the selection of candidate genes or regions, where gene prioritization and genetic variant selection tools can be applied [36]. With the advancement of next-generation sequencing (NGS) regions can also be selected. This selection can incorporate disease or domain specific aspects, such as preferences for specific control mechanisms (e.g., miRNA-SNPs) and population specific filters as well [37]. Although the application of multiple, weakly dependent phenotypic descriptors, describing independent aspects of the phenotypic target concept can substantially increase power, general dichotomous case–control sample size considerations are advised. A practical reason is that standard statistical tools are typically applied in parallel [38]. These analysis and meta-analysis aspects should also be considered, because CGASs are increasingly applied as confirmatory follow-up studies in multistage design. The selection of phenotypic variables is largely determined by the available clinical endpoints and endophenotypes, but the inclusion of potential confounders and covariates for adjustment is essential as effect strength estimation is typically an important goal in these studies. In this association analysis and Bayesian statistical context, the general theoretical sample size considerations applied in PGM learning are not applicable [39]. Case studies, systematic power and sample size evaluations using simulated data and real-world data sets for PGMs in CGAS and GWAS were reported in [11, 14, 15, 19–22, 40]. These studies, in line with our experience, confirm that this scenario of Bayesian exploration of relevant variables using PGMs complies with the "rule of ten" samples per variable, known in other multivariate conditional modeling approaches, such as in logistic regression [41].

3 Materials

3.1 Data

In this chapter we use a rheumatoid arthritis (RA) data set for illustrational purposes (for details *see* **Note 1**). Investigated SNPs included the KLOTHO1 G-395A (promoter, rs1207568), the KLOTHO2 C1818T (exon 4, rs564481), GUSB1 (rs709607), GUSB2 (rs906134), exonic SNP GUSB3 (rs9530), SNP near the 3′ region GUSB4 (rs12538430), and an intronic SNP GUSB5 (rs1880556). The study population consisted of ($n = 167$) patients suffering from RA and ($n = 302$) healthy controls. The following parameters of the patients with RA have been recorded and evaluated during the study for correlation with SNP alleles: presence of rheumatoid factor (RF), C-reactive protein (CRP) level, erythrocyte sedimentation rate (ESR/WE), disease activity score 28 (DAS 28 score), age of the patient, age of disease onset, blood pressure, presence of electrocardiogram (ECG) alterations, hypertension, myocardial infarction (MI), levels of serum cholesterol, triglyceride, uric acid, fasting blood sugar. Coronary artery disease is a frequent comorbidity in RA, and SNPs of the KLOTHO gene were found to be associated with coronary artery disease, this is why several laboratory parameters, such as serum cholesterol and triglyceride levels were determined.

3.2 Computation

- The computational background is available at http://bioinformatics.mit.bme.hu/tools/bmla after registration.

3.3 Postprocessing and Visualization

- We illustrate the BMLA methodology through a case study in rheumatoid arthritis (RA) using the RA data set and BayesEye, our free postprocessing and visualization tool. BayesEye is available at http://bioinformatics.mit.bme.hu/tools/bayes-eye. Various precomputed statistics for the BMLA methodology from the RA data set are available at http://bioinformatics.mit.bme.hu/tools/bmla.

3.4 Gene Prioritization

- The gene prioritizer tool applied after BMLA can be accessed at http://bioinformatics.mit.bme.hu/tools/GP.

4 Methods

4.1 Filtering and Imputing SNP Data

We start processing the data set by investigating errors (i.e., measurement errors marked in the data set). Since the Bayesian relevance analysis requires a complete data set, we need to define two thresholds for the ratio of missing data. In this case, all improper data cells of the data set (such as erroneous cells and missing values) count as missing. The first ratio should define the acceptable missing data rate per variable (MRV), i.e., a column of the data set, e.g., SNP, and the other threshold should define the missing data rate per sample (MRS).

1. The first step is discarding (nearly) totally missing SNPs (MRV > 95 %).
2. Then the samples should be filtered (i.e., removed in case of MRS above the selected threshold) depending on the data set size and sample quality (*see* **Note 2**).
3. Variables with high MRV should be discarded (*see* **Note 3**).
4. Furthermore, the variability of variables should be also taken into account. Variables with a single possible value, e.g., monomorphic SNPs, must be discarded (*see* **Note 4**). More generally, a variable with variability less than 1 % (i.e., if a value of the variable appears in less than 1 % of the data or in less than 10 cases) should be removed, because nearly constant values can cause anomalies in BN learning.
5. The next step is the testing of Hardy–Weinberg equilibrium (HWE) for all SNPs (*see* **Note 5**). SNP that have a significant *p*-value for the test of disequilibrium in controls should be discarded as this generally indicates a measurement error.
6. In many real world applications the incompleteness of the data poses a further challenge apart from facilitating inference. There are several software tools that offer various imputation methods. Some are tailored especially to the imputation of SNPs (*see* **Note 6**).

(a) If a SNP has no related SNPs with which it is correlated, then a simple imputation-based on the unconditional distribution of the SNP may be sufficient.

(b) On the other hand, in case of strong correlations more complex methods, such as the one implemented in IMPUTE, should be used.

7. To focus on the complete BMLA workflow, we assume a complete data set (*see* **Note 7**). Bayesian relevance analysis does not allow the explicit modeling of unobserved or in other words hidden variables. Therefore, such a variable has to be discarded or imputed based on some reference or other correlated variables.

4.2 Haplotype Reconstruction

Haplotype reconstruction creates new haplotype block variables based on individual SNP variables (*see* **Note 8**). The cardinality of these variables is greatly increased. Instead of three (AA, Aa, and aa) it can be well over 100. Most methods cannot handle this degree of cardinality, especially with many of the haplotypes being rare. In order to perform a meaningful analysis the cardinality must be reduced with minimal loss of information.

- The simplest approach is *sample doubling* which halves the cardinality. In this case, we simply take the first haplotype from each genetic region, assign it the same target variable value and add it to our samples, and repeat with the second haplotype from each sample (*see* **Note 9**).
- *Variable doubling* or *diplotype construction* is another approach, in which samples are separated into two variables, one for each strand (*see* **Note 10**).
- A further, haplotype frequency-based approach selects a haplotype for a given sample based on the selected *disease model*. In case of a dominant disease model the more frequent haplotype is selected for each sample, whereas in the opposite case of a recessive disease model the less frequent haplotype is used.
- *Aggregating rare haplotypes* may also reduce cardinality significantly. This method can be applied in conjunction with either of the previous methods. Using a predefined minimal haplotype frequency, rare haplotypes are merged to frequent ones based on their base-pair distance (hamming distance). If multiple haplotypes at equal hamming distance exist, the more frequent one is chosen.
- There are also alternative, non-haplotype-based aggregation methods for SNPs that can be applied instead (*see* **Note 11**).

4.3 Knowledge-Rich Aggregation

Through "positional" or "functional" integration, SNP level results can be aggregated to higher abstraction levels such as genes, proteins, and pathways. This aggregation can take place either in the preprocessing phase by creating new synthetic variables to be used in the data analysis, or in the interpretational (postprocessing) phase (*see* **Note 12**).

Given a taxonomy file detailing the relationships for each SNP (e.g., SNP–gene relationships), BMLA can facilitate an appropriate aggregation of results.

4.4 Phenotype Data Transformation and Discretization

Further processing of the data may be required, depending on the available phenotype descriptors, clinical and environmental variables. Contrary to genetic data, phenotypes and other descriptors cannot be imputed. Therefore, the preprocessing of these variables can be crucial for a successful analysis.

1. In case of multiple quantitative phenotype descriptors that may serve as "targets" of interest, there is a conventional approach of merging and transforming these variables into a complex phenotype (*see* **Note 13**). Such merging of targets is still an option within the Bayesian framework, but it is not necessary, since the Bayesian relevance analysis can handle multiple targets at a time. That is, instead of a "black box" merging, the relationships of all investigated targets can be analyzed.
2. The current BN-BMLA framework used for the relevance analysis assumes discrete variables, and therefore, all quantitative phenotypes, environmental and clinical factors have to be discretized (i.e., binned). There are several basic methods to perform binning, such as the equal bin width method provided by BayesEye.

4.5 Prior Selection

The Bayesian framework allows the incorporation of prior knowledge (*see* **Note 14**) on multiple levels in the forms of hard (constraint-based) and soft (quantitative) structure priors and parameter priors. Finding an appropriate combination of prior parameters and settings that match the application domain requires the joint consideration of hard, soft, and parameter priors (*see* **Note 15**).

1. *Hard structure prior.* The search in DAG space can be hastened by imposing reasonable restrictions on DAG structures, i.e., on model complexity. Limiting the number of incoming edges into a node (i.e., the number of parents) is a frequently used policy to avoid unrealistic or non-preferable structures. This limit should depend on the number of variables and the sample size.
   (a) In cases where the sample size is small compared to the number of variables (i.e., sample size should be an order of magnitude higher than the number of variables) a lower threshold such as 2 or 3 should be selected (*see* **Note 16**).
   (b) At the other end of the scale, in case of an adequately high sample size, a larger threshold, e.g., 7 or 8, may be used. The high computational cost may prohibit the use of larger thresholds.

2. *Soft structure prior.* If domain specific knowledge is available, then a higher level structural prior can be defined for certain structural features. This might mean that specific connections between variables are favored or even that complex subgraphs are predefined. Further soft priors can be defined regarding the probability of sets of incoming edges, i.e., parental sets (*see* **Note 17**). Otherwise, when no domain knowledge is available, a uniform structure prior should be used. However, even in such a case the prior edge probability (for edges connected to the target) should be defined based on an assumption concerning the number of possibly relevant SNPs (e.g., in GWAS 1 in 10^4, or even 1 in 10^6).
   (a) In CGAS/PGAS studies, as these SNPs are typically selected based on previously defined criteria, a relatively large prior edge probability of 0.1 (1 in 10) can be used.
3. *Parameter priors* influence the likelihood scoring function, thus they play an essential role in Bayesian learning. The Bayesian Dirichlet (BD) prior is a popular choice, which we also recommend (*see* **Note 18**). Regarding hyperparameters:
   (a) for small, moderate and adequate sample sizes (with respect to the number of variables) use Cooper–Herskovits hyperparameters,
   (b) for asymptotically large sample sizes use BDeu hyperparameters (*see* **Note 19**).

4.6 MCMC Simulation in BN-BMLA

Due to the high computational complexity of computing posteriors for the structural properties of Bayesian networks we used a Markov chain Monte Carlo (MCMC) method, which facilitates a random walk in the space of directed acyclic graph (DAG) structures (*see* **Note 20**).

1. *Simulation length.* The sufficient length of the MCMC sampling process depends on multiple factors, and it cannot be determined analytically. Therefore the application of MCMC methods is guided by various convergence diagnostics (*see* **Note 21**).
   (a) We found that a burn-in length of 10^6 steps and a sample collection length of 5×10^6 steps are sufficient in most scenarios in which there are at most 100 variables (*see* **Note 22**).
2. *Semantic settings.* Structural properties of Bayesian networks provide a rich language that can be used to form queries, which are evaluated using the basic posteriors (*see* **Note 23**) estimated via the MCMC process. Example queries:
   (a) Markov blanket membership (MBM) probabilities (*see* subsequent section and **Note 24**) of all variables.
   (b) Probability of single edges based on the DAG structures traversed by the MCMC process (*see* **Note 25**).
   (c) Causal relations (*see* **Note 26**).

4.7 Postprocessing and Visualization of BMLA Results

We provide a downloadable GUI (BayesEye) for the analysis and visualization of the BN-BMLA results (*see* **Note 27**).

1. *MBM visualization.* Markov blanket membership (MBM) is a structural property of Bayesian networks (*see* **Note 23**), which entails strong relevance. In the BMLA methodology within the Bayesian statistical framework, it provides the most aggregated view of multivariate relevance in the form of univariate posterior probabilities, i.e., values are in the range of 0 (lowest probability) to 1 (highest probability). The posterior of the strong relevance of a given variable means the probability that the statistical association of the variable is direct, i.e., not mediated transitively by other variables analyzed in the study (formally, it is in the Markov blanket of a selected target). Illustrational results are shown in Table 1 for the RA case study (*see* **Note 28**).
2. *MBS visualization.* Markov blanket sets (MBS) provide a multivariate characterization of strong relevance. Each MBS contains various numbers of members (variables). The posterior of an MBS represents the relevance of that set of variables (*see* **Note 29**).
   (a) Model averaging is highly recommended as selecting only the *maximum a posteriori* could lead to improper results (*see* **Note 30**).
   (b) Especially in case of a refined analysis, exploring weak and conditional relevance for detailed phenotyping, Bayesian model averaging is necessary.
3. *Refined characterization of relevance for sets of predictors.* Additional processing steps may be necessary in order to help the interpretation and to identify the most relevant results. The following tools are optional extensions of MBS analysis, each providing a different characteristic.

Table 1
Posterior probabilities of strong relevance

MBM	RA	RA+	Multitarget
gusb1	0.2032	0.1412	0.1443
gusb2	0.8795	0.6904	0.7174
gusb3	1.0000	0.5597	1.0000
gusb4	0.0038	0.0008	0.0009
gusb5	1.0000	0.5628	1.0000
klotho 1	0.0399	0.0352	0.0364
klotho 2	0.0977	0.0811	0.0867
Gender	0.5473	0.5249	0.6203

(a) *Conditional relevance.* The BayesEye software provides tools for visualizing multiple MBSs at a time (by means of model averaging), and also for the filtering of MBSs by a specific condition, e.g., to show their conditional posterior probabilities. A condition (filter) can be created as a Boolean expression of variables.

(b) *Relevance map.* The reason behind this concept is that even in case of several MBSs with equally flat posteriors there can be common parts (stable patterns) that are present in most sets (*see* **Note 31**). A k-subMBS contains variables that are strongly relevant with high confidence ("necessary" variables), and the complement of a k-supMBS contains variables that are not strongly relevant with high confidence (i.e., k-supMBS contains a "sufficient" set of variables).

(c) *Relevance tree.* A further tool for visualizing k-subMBS results is the so called relevance tree (*see* **Note 32**). Subsets with posteriors larger than a predefined threshold are visualized in a tree structure.

(d) *Interactions view.* The *interaction-redundancy score* explicitly quantifies whether a variable is part of a given subset due to an interaction, or because it is relevant independently of others (*see* **Note 33**). Based on this score, interactions and redundant relationships can be visualized.

4. *MBG visualization.* MBGs provide detailed information on the relationships of relevant variables with respect to the target(s). They allow the investigation of relationship types, whether a variable is directly related to a target or indirectly through an interaction (*see* **Note 34**). The BayesEye software also provides a tool for averaging and investigating MBGs, similarly as for strong relevance (MBSs). A condition (filter) can be created using a Boolean expression of relationships (edges) or variables (nodes), which allows focusing on various types of MBGs (e.g., to see their conditional posterior probabilities).

   (a) Averaging of MBGs is recommended as investigating the *maximum a posteriori* MBG only leads to incomplete results in most cases.

   (b) In case of large MBGs, focusing on the most relevant parts by filtering provides a clearer view of the results.

4.8 Bayesian Network-Based Structure Conditional Effect Strength Estimation

The model-based approach using Bayesian networks allows a wide variety of effect strength concepts, such as effect size measures conditioned on structural properties, e.g., Markov blankets. These methods perform full Bayesian model averaging, both at structural and parametric levels, thus providing special Bayesian odds ratio measures (*see* **Note 35**).

1. *Markov blanket conditioned odds ratio.* MBGs can be used for the estimation of the posterior distribution of odds ratios, since they act as a filter containing only those dependency relationships of variables in which the target is closely involved (*see* **Note 36**). An MBG-based odds ratio is a hybrid measure providing a dual view of relevance including both structural and parametric aspects.
2. *Credible interval.* Based on the posterior distribution of an MBG-based odds ratio, a high probability density region can be identified, which allows the computation of a credible interval (i.e., the Bayesian equivalent of a confidence interval).

4.9 Reintegration of CGAS Results to Omic Levels by Gene Prioritization

For the omic extension of the CGAS results, a special data and knowledge fusion methodology, called computational gene prioritization (GP), can be applied. Gene prioritization (GP) takes a gene set as search terms ("query") and based on the coherence of this set, it determines a fully automated weighting of information sources and produces a full (or partial) ranking of genes (from the "input gene list") indicating relevance to the query. Genes that are more similar to the query genes are ranked higher, and genes dissimilar to the query are ranked lower in the resulting prioritization (*see* **Note 37**).

1. *CGAS-based input list construction.* The hypothesis driven design of a CGAS provides multiple ways to select the genes that we want to prioritize, i.e., the input gene list.
   (a) In the most restrictive case, the input list is constrained to the genes used in the CGAS/PGSS. In this scenario, the subsequent GP after the data analysis combines the results of data analysis and omic knowledge, i.e., the overall workflow combines CGAS data and prior omic knowledge.
   (b) In a more relaxed scenario, we can expand the set of CGAS genes or relevant genes from CGAS based on a former genome-wide association analysis, in essence reversing the design process of the CGAS, i.e., in this case the most promising genes based on earlier GWASs and the CGAS can compose the input list.
   (c) In the most general case, the whole genome can be used as an input list, although GP methods originally were designed for input lists with a few hundred items.
2. *Selection of omic information sources.* We have to select appropriate omic information sources in order to integrate background knowledge with the results of our analyses.
   (a) We chose STRING and GEO databases (*see* **Note 38**).

3. *Disease specific kernel construction.* To create a gene–gene similarity kernel, we can calculate, among others, a normalized Laplacian of the weighted network of the corresponding protein–protein interactions (PPIs) (*see* **Notes 39** and **40**). We show an example of how to incorporate disease-specific gene expression information into the PPI kernel.
   (a) First, we search gene expression data sets in which patients with RA were investigated in the Gene Expression Omnibus (e.g., GSE1919, GSE7669).
   (b) Next, we download the series matrix file of the experiments (the normalized expression data of the samples by probe identifiers).
   (c) Then aggregate the expression level of the probe identifiers to the level of genes (by calculating the mean expression level of the probe identifiers related to a given gene), and calculate the Pearson correlation coefficients of all gene pairs. This process results in one valid gene expression-based kernel matrix per GEO experiment.
   (d) In order to integrate these disease-specific kernels with the PPI kernel, we calculate the element-wise product of the gene expression-based kernels and the PPI kernel matrices. The resulting combined kernels can be seen as disease-specific PPI kernels. Finally, we can use these kernels in the gene prioritization process.
4. *CGAS-based query construction.* If GP is meant to support the CGAS result interpretation, query construction becomes further complicated by the CGAS design (*see* **Note 41**). For example, using a query consisting of genes found weakly relevant by the CGAS, the results of GP can reflect the CGAS design principles rather than the content of CGAS data (*see* **Note 42**). It is advised to compose multiple queries consisting of genes with varying degrees of relevance on the basis of the CGAS. Additionally, these queries may also be augmented with genes based on prior knowledge.
5. *Interpretation and quality assessment.* The output of the prioritization is a sorted list of the input genes. If we prioritized the whole genome (or high number of genes) then the manual investigation of the results can be very tedious. In addition, because of the complex nature of the kernel fusion-based gene prioritization method, explaining the contribution of the snippets of similarity information to the final ranking is a difficult task. Additional tools can be utilized to support data analysis and interpretation:
   (a) *Network analysis.* Similarity networks can be constructed by representing top-ranking genes in a network, with the edges labeled with the combined similarity of the genes.

This network can aid the interpretation of the results by showing which query genes are similar to the top-ranked genes and how much they are similar to each other (*see* **Note 43**).

(b) *Enrichment analysis.* To uncover the hidden patterns embedded in the ensemble of ordered entities, enrichment analysis can be utilized to estimate whether a predefined set of genes are overrepresented among the high-ranking entities (*see* **Note 44**).

(c) *Filtering.* Results can be filtered out on the basis of various properties including gene function, previously known disease genes (if we are interested in novel findings), or on the contrary: novel genes (for validation purposes).

4.10 Semantic Publishing of GAS Results

The documentation of previous steps can be standardized according to the recommendations for reporting genetic association studies (for a summary, with special emphasis on meta-analysis and dissemination of the weakly significant results, *see* **Note 45**).

4.11 Concluding Remarks

Detailed phenotypic descriptors and rare genetic variants pose a dual challenge for association research. To achieve this, hypothesis-driven candidate gene association studies can efficiently complement omic level studies, because the decreased number of analyzed factors in CGAS allows the application of algorithms which are not tractable at higher omic levels, such as methods using Bayesian networks for effect strength estimation and causal analysis. The discussed workflow using the Bayesian network-based Bayesian multilevel analysis combined with gene prioritization illustrates a possible interplay between hypothesis-free and hypothesis-driven approaches.

5 Notes

1. *RA study background*: Earlier works indicated a wide range of genetic and omic markers related to RA (for a causal approach *see* for example [10]; for a Bayesian approach, *see* [42]; for prognostic factors, *see* [43]). Earlier studies showed that β-glucuronidase, present in the synovial fluid of RA patients, may contribute to the depletion of glycosaminoglycans from cartilage allowing invasion of synovial cells [44]. In a later work the expression of β-glucuronidase-encoding genes, such as the anti-aging KLOTHO and the lysosomal hydrolase GusB was examined in patients suffering from RA [45].
2. *Sample filtering*: In order to concentrate on the whole workflow, we do not discuss the utilization of quality and uncertainty information from genotyping. We assume a standard coding either with binary variables in dominant/recessive models or tertiary

variables (e.g., 0-1-2, where 0 is a wild-type homozygote, 1 is a heterozygote, and 2 is a mutant homozygote).

Note that due to the multivariate analysis, missing values in a sample can either be imputed or the sample is discarded. A strict threshold can be selected on a large data set with high quality samples, such as MRS: 5–10 %. In more practical scenarios we may select a more permissive threshold such as MRS: 20–25 %. However, with low sample size and moderate sample quality MRS may be as high as 50 %. Note that samples with missing values for key target variables should be discarded regardless of their MRS.

3. *Variable filtering*: This threshold may also depend on the quality of the data set. We consider 5 %, 10 %, and 20 % as strict, average, and permissive thresholds, respectively.
4. *Variable variability*: For example in the RA study case, an additional GUSB SNP had to be removed, because it was monomorphic having only common homozygous cases.
5. *Hardy–Weinberg equilibrium analysis*: Among the several tools for testing HWE, the online tool of Wigginton et al. [46] is the most popular choice. Note that apart from HWE, this online tool provides several other frequentist association test results.
6. *Imputation tools*: There are several software programs that facilitate imputation. Complex statistical software programs such as SPSS have a corresponding imputation module. Other tools such as IMPUTE [47] are designed especially for the imputation of SNPs.
7. *Imputation*: The reasons for missing can be numerous, a typical case is that not all variables are analyzed in each sample due to insufficient resources or to improper study protocol (e.g., obfuscated clinical variables). Another typical reason is that the value of a certain variable cannot be identified unambiguously.

   There are several approaches to handle missing values in case of Bayesian network learning. The Expectation-Maximization (EM) algorithm is one of the most well-known methods for learning parameters of statistical models from incomplete data [48]. Another approach called Data Augmentation was based on stochastic simulation method, and was devised by Tanner and Wong [49].

   Note that the general assumption about missing data is that it obeys the Missing at Random (MAR) mechanism [50]. Basically, MAR means that the probability of a missing value for a variable of interest only depends on other observable variables. In case the MAR assumption is violated due to systematic missing, additional processing steps are needed.

8. *Haplotype reconstruction*: The central concept in converting SNPs to haplotypes is centered on exploiting the linkage disequilibrium that exists between close loci to transform sets of SNPs to haplotypes. The sets are most often defined by the genes that the SNPs are located on. In case of large genomic regions spanning more than 50 kb, multiple haplotype blocks can be defined, thus allowing separation of non-tightly linked regions. Haplotype blocks can be identified with software like Haploview [51], other tools allow the reconstruction and a subsequent analysis of haplotypes, such as HapScope [52], HAPLOT [53], GEVALT [54], PHASE [55]. In the RA case study and in several previous studies, we used PHASE.
9. *Sample doubling*: An advantage of this method is simplicity, while a large disadvantage is the fact that the order of the first and second haplotypes are not completely random, and they do NOT correspond to the maternal or paternal strand. Thus, a sampling mix of the inputs must be produced.
10. *Variable doubling*: In case of using PHASE to reconstruct haplotypes, the characteristics indicate that the more frequent haplotype will be placed on strand 1 (first in the list) and the rarer one will be placed on strand 2. H1 and H2 correspond to each strand.
11. *Alternative SNP aggregation: clustering*: Another approach towards aggregating SNPs is to form clusters of SNPs based on their function and location information. This method allows to create complex variables that represent the elements of a cluster [56].
12. *Knowledge rich aggregation*: The haplotype level analysis is a well-known example for facilitating the aggregation in the preprocessing phase. In case of a frequentist framework, the most commonly used method for aggregation in the postprocessing phase is the combination of p-values for SNPs into an overall significance level to represent the association of the corresponding gene or pathway with the disease [57]. However, these combination methods are vulnerable to correlations between the entities, and they cannot cope with complex interactions.

    The Bayesian framework using Bayesian networks offers a principled solution for the multivariate propagation of uncertainties. Since the model posterior can be transformed and interpreted without theoretical restrictions, given the space of Bayesian network structures, the posterior can be aggregated by any partitioning over model structures G. Beside noninformative model aggregation, the prior domain knowledge can be used to create meaningful partitions, e.g., genes from SNPs.
13. *Phenotype data transformation*: In a frequentist framework, one typically carries out as many separate analyses as the number of chosen targets, which are typically disease or

phenotype descriptors. This may result in overly stringent p-value thresholds (i.e., correction) for association tests, which can be avoided by an appropriate selection and transformation of target variables. For quantitative target variables, clustering and dimensionality reduction, such as principle component analysis are popular choices for identifying relevant phenotype components and for forming a composite phenotype to be used for the association analyses [58]. Regarding discretization, there are several solutions such as sophisticated, general BN-based methods [59].

14. *The Bayesian approach*: The essence of Bayesian methods is the combination of a priori probability distribution and a likelihood, resulting in the a posteriori probability according to the Bayes' theorem. A Bayesian network $BN(G, \theta)$ consists of a directed acyclic graph structure G and its parameterization θ. Variables (factors) and their dependency relationships are represented in a graph structure by nodes and directed edges respectively. Dependency relationships between the modeled variables are characterized by conditional probability distributions denoted as θ (parameterization).

    In case of Bayesian network structure learning, given a complete data set D and a directed acyclic graph G that represents the joint probability distribution of discrete random variables $X = X_1, \ldots, X_n$ having a multinomial distribution, the aim is to estimate the posterior $P(G|D) \sim P(G)P(D|G)$. In order to get the posterior of a certain structure given the data, a prior distribution of the possible structures $P(G)$ and a likelihood $P(D|G)$ are needed. Based on this relation the posterior probability $P(f|D)$ of any structural property of Bayesian networks can be estimated as

    $$P(f \mid D) = \sum_{G} P(G \mid D) f(G)$$

    where $f(G)$ is 1 if the property holds in G and 0 otherwise.

15. *Prior selection*: Priors (a priori probability distributions) are fundamental technical elements of the Bayesian framework, which also allow the incorporation of a priori knowledge, e.g., conclusions from previous studies. If there is no specific domain knowledge that requires prior incorporation, non-informative priors should be used.

    Previous works described the application of priors consisting of a mixture of normal distributions for univariate Bayesian methods used in the GAS analysis of SNPs [24]. These normal priors required a predefined value for the proportion of SNPs having a non-zero effect size, which was to be determined by the investigator within the recommended range of 10^{-4} to 10^{-6}. Contrary to the conventional hypothesis testing framework, this value is not dependent on the number of tests carried out, and no further correction is required due to multiple hypothesis testing. The correction is internal because the posterior takes the weighted average of the possible models into account. However, the monitoring of the false discovery rate is also recommended. The problem with this univariate approach is that it considers SNPs as independent entities, thus excluding the possibility of higher dependency patterns.

In a multivariate Bayesian framework a connection was established between the virtual sample size (VSS) hyperparameter of the non-informative Dirichlet prior and an effect strength measure in the form of log-odds. That is the setting of the VSS defines the probability distribution of effect sizes [60]. The selection of a structure prior (i.e., probability of dependency relationships between variables) is also required in a multivariate framework similarly to the a priori ratio of relevant SNPs in the univariate case, which is specified in two parts, as model complexity constraints (hard structure prior) and quantitative beliefs (soft structure priors).

16. *Hard structure priors*: Selecting an overly strict threshold such as 2 may seriously impact the capability of discovering interactions. Therefore, it should only be used when the sample size is rather small. A fail-safe approach is to run the computations with multiple thresholds and compare results. Many variables becoming equally relevant can be a sign for an improperly high threshold. In the RA case study we used 2 and 3 as thresholds with a burn-in of 10^6 steps and sample collection of 5×10^6 steps.

17. *Soft structure priors*: Within the BN-based BMLA framework, various soft priors can be defined regarding the probability of sets of incoming edges (i.e., over parental sets). Possible options include:

    (a) Uniform probability for sets of incoming edges with a size below a predefined threshold [61],

    (b) Uniform probability over set sizes, and uniform probability for sets having the same size [33],

    (c) Predefined prior edge probability (assuming independence of edges) [61].

18. *Parameter priors*: The selection of an appropriate prior (i.e., a suitable type with adequate hyperparameters) for Bayesian methods is a well-known problem. Even though there are previously reported empirical results [62–64], there are no general methodologies for prior selection.

    The Bayesian Dirichlet prior is a popular choice, since it assigns the highest score to the maximum a posteriori structure, and it is also a conjugate prior for multinomial sampling [65]. Being conjugate means that the distribution function of the prior and the posterior belongs to the same family under a given sampling model, which is an essential property, since it enables analytical computation.

19. *Hyperparameters*: Parameters that allow a refined tuning of prior distributions, such as Bayesian Dirichlet, are called hyperparameters in general. In the RA case study we used two types of Bayesian Dirichlet parameter priors, the first is defined by the Cooper–Herskovits hyperparameters [61], and the second is defined by the BDeu hyperparameters with virtual sample size of VSS = 1 [62].

20. *MCMC*: This directed acyclic graph (DAG)-based process allows the estimation of all the structural properties of Bayesian networks, assuming a complete data set DN, discrete variables, multinomial sampling, Dirichlet parameter priors, and a uniform structure prior [15]. These assumptions allow the derivation of an efficiently computable unnormalized posterior over the Bayesian network structures (i.e., over DAGs) [61, 66].

    Using this unnormalized posterior, the MCMC method performs a special random walk in the space of DAGs by applying operators for inserting, deleting, and inverting edges [67]. The probability of applying these operators in the proposal distribution is uniform. In each MCMC step, the Markov blanket graph (*see* **Note 18** and [16]) corresponding to the DAG in the current step is determined and the relative frequency of this Markov blanket graph is updated.

21. *Convergence diagnostics*: There are various quantitative measures for assessing the convergence of Markov chains and to determine the confidence intervals of the posteriors. The following set of measures can be regarded as a standard set: (1) The Geweke *Z*-score: measures convergence within one chain, i.e., the significance of the difference of the posteriors between the beginning and the end of the sampling [50]. (2) The Gelman-Rubin *R*-score: measures inter-chain convergence, i.e., significance of the difference of independent sampling processes [50]. (3) Confidence intervals, based on the standard error of the MCMC sampling calculated according to the batch means method [68].

22. *Burn-in*: In an optimal scenario, the MCMC process enters into a stationary state after a sufficient number of steps, in other words convergence is established. The period required to reach this state is called the burn-in phase, during which all samples are discarded. The run length depends on the complexity of the model, i.e., the number of variables and the size of the data set.

23. *Markov blanket sets and graphs*: Learning a Bayesian network structure *G* means finding a DAG that best represents the data set. However, in most cases, the sample size of the data set is not sufficient with respect to the number of variables to select

a single best model. Rather, there are several models with non-negligible posteriors. This means that learning G can be extremely challenging if not impossible. A possible solution is to use Bayesian model averaging and less complex structural properties. Even if the learning of the whole model is not feasible, certain structural features can be extracted reliably. A Markov Blanket Set (MBS) is a central structural feature, which is related to the concept of strong relevance of a single variable or a set of variables.

Definition—Markov blanket set: A set of variables X' is called a Markov Blanket Set of Y if conditioned on X' the "target" variable Y is independent of all other variables. Formally, given the set $\mathbf{V}$ containing all variables $X_1, \ldots, X_n$, a subset of variables $\mathbf{X}' \leq \mathbf{V}$ is a Markov blanket set of Y with respect to the distribution $p(X_1, \ldots, X_n, Y)$ if $Indep(Y, (V \backslash X') | X')$, where $Indep()$ denotes conditional independence [69].

The reason for learning the MBS of a target variable Y is that given the stability condition and the faithfulness condition [35] the elements of this set are strongly relevant with respect to Y.

Definition—strong and weak relevance: A feature X_i is strongly relevant to Y, if some $X_i = x_i$, $Y = y$, and $s_i = x_1, \ldots, x_{i-1}, x_{i+1}, \ldots, x_n$ for which $p(x_i, s_i) > 0$ exists, such that $p(y|x_i, s_i) \neq p(y|s_i)$. A feature X_i is weakly relevant, if it is not strongly relevant, and a subset of features S_i' of S_i exists for which some x_i, y, and s_i' exist for which $p(x_i, s_i') > 0$ such that $p(y|x_i, s_i') \neq p(y|s_i')$. A feature is relevant, if it is either weakly or strongly relevant; otherwise it is irrelevant [70].

In other words, the MBS provides a proper multivariate characterization indicating the strong relevance of a set of variables given the stability condition [71]. A further related structural property is called the Markov Blanket Graph (MBG), which in addition to the nodes of an MBS comprises of the dependency relationships of corresponding variables represented as directed edges.

Definition—Markov blanket graph: An MBG(Y,G) of a variable Y is a subgraph of a Bayesian network structure G, which contains the nodes of the Markov blanket set of Y, that is MBS(Y,G) and the incoming edges into Y and its children. Given a target node, which corresponds to the target variable Y, MBG(Y,G) as a (sub)graph structure consists of nodes that are (1) parents of Y, (2) children of Y, or (3) "other parents" of the children of Y.

24. *Markov blanket membership*: MBM is a univariate aggregate of the multivariate MBS posteriors, and allows the assessment of the relevance of individual variables [33],

    Definition—Markov blanket membership: The Markov Blanket Membership MBM(X_i,Y) holds if $X_i \in \mathrm{MBS}_Y$.

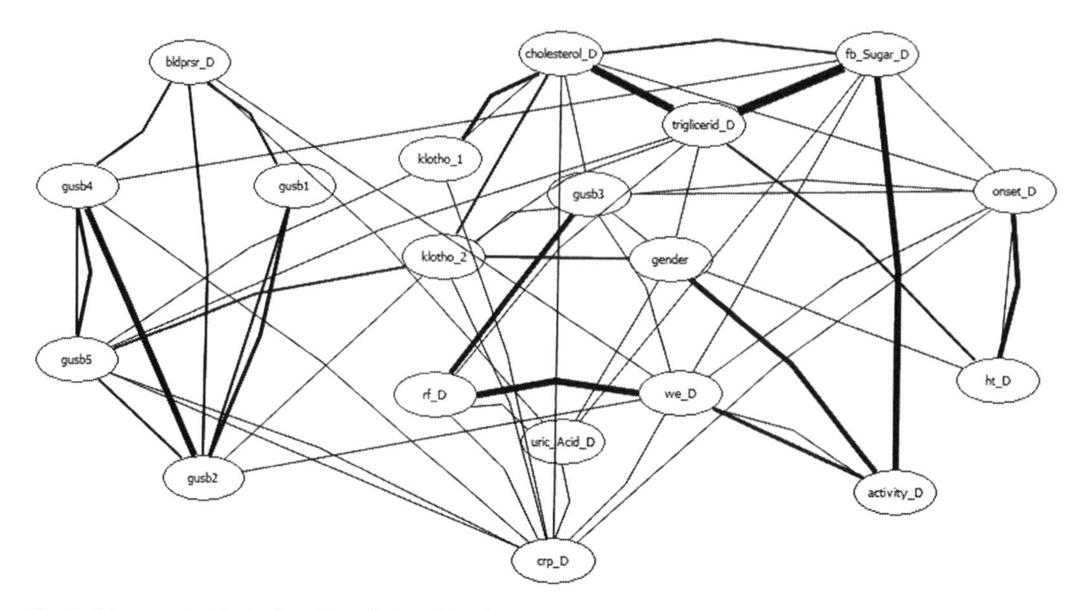

Fig. 1 Edge graph illustrating the relationships between variables of the RA case study based on the DAG structures traversed by the MCMC process

25. *Edges*: It is also possible to investigate the probability of single edges based on the DAG structures traversed by the MCMC process. Figure 1 shows the results of the RA case study, where edges have various widths according to their probability.

26. *Causal relations*: Structural properties can also be used to infer certain types of causal relations, e.g., whether a possible causal relationship between two factors is directed or transitive. A direct causal relationship with the target either means that target is directly influenced by or the target directly affects a certain variable. So there are no other variables mediating the relationship in a direct connection, whereas in a transitive relationship there are.

27. *Visualization*: BayesEye is also capable of visualizing the results of non BN-based Bayesian multivariate association analyses, e.g., using logistic regression or decision trees [72, 73].

28. *Interpretation of MBM analysis*: Table 1 contains strong relevance posteriors with respect to RA given genetic factors (GUSB, KLOTHO) and clinical factors ("activity"—disease severity, "onsetAge"—disease onset). In case of the RA data set only genetic factors and gender were investigated, whereas RA+ contained the clinical factors as well. In both cases RA served as target. The Multitarget data set was the same as RA+, but "onset Age" and "activity" were treated as targets (i.e., the model had multiple targets). In general, posteriors over 0.9 may be considered as "highly relevant," between 0.7 and 0.9 as "relevant," 0.5 and 0.7 as "moderately relevant."

Lower posteriors, such as 0.3–0.5 should only be considered somewhat relevant in case of small sample size [24]. Results indicate that GUSB3 and GUSB5 are highly relevant in the RA and Multitarget cases with a maximum posterior of 1. GUSB2 also appears to be relevant, with high posteriors (0.8795 and 0.7174 respectively), whereas gender is moderately relevant.

In the RA+case the clinical factors "activity" and "onsetAge" are highly relevant with respect to RA (with posteriors 0.958 and 0.90 respectively) which is expected as they are closely related. The interesting phenomenon is that they have a push out effect on the genetic factors causing their posteriors to decrease, although they still can be considered as relevant. In other scenarios however, this phenomenon can be so severe that phenotype descriptors and clinical factors can completely shield off the effect of genetic factors resulting in negligible posteriors.

29. *MBS complexity*: The main difficulty of MBS analysis is its complexity (i.e., the number of possible sets), which can be reduced by filtering and marginalizing. More specifically, marginalizing out insignificant variables (i.e., those of near-zero MBM posteriors) reduces MBS space while only a negligible loss of information occurs.

30. *MBS averaging*: In practical scenarios the MBS distribution is more or less flat in the sense that there are many similar sets with similar posteriors. So a relatively low posterior does not necessarily mean irrelevance, rather that there are many MBSs that are similarly relevant. For example in the RA case study the maximum a posteriori MBS had a posterior of 0.3. Therefore, investigating several sets and identifying common members is crucial. The BayesEye software provides tools for visualizing multiple MBSs at a time (by means of model averaging). Furthermore, it allows the filtering of the MBS for a Boolean statement of variables.

31. *Relevance map based on k-subMBS and k-supMBS subsets*: For a distribution $p(V)$ with Markov Boundary set mbs, a set s is called sub-relevant if it is a k-ary Markov Boundary subset (k-subMBS), i.e., $|s| = k$ and $s \subseteq mbs$. A set s is called sup-relevant if it is a k-ary Markov Boundary superset (k-supMBS), i.e., $|s| = k$ and $mbs \subseteq s$. Figure 2 shows a heatmap and a graph detailing possible k-subMBS and k-supMBS subsets of various sizes created by the BayesEye software for the RA case study.

    Investigating the most probable k-subMBS and k-supMBS sets of various sizes holds valuable information on the patterns and interactions of relevant variables. The so called relevance map is a visualization aid for this type of analysis, where each node in the map represents a given set of predictors, e.g., the

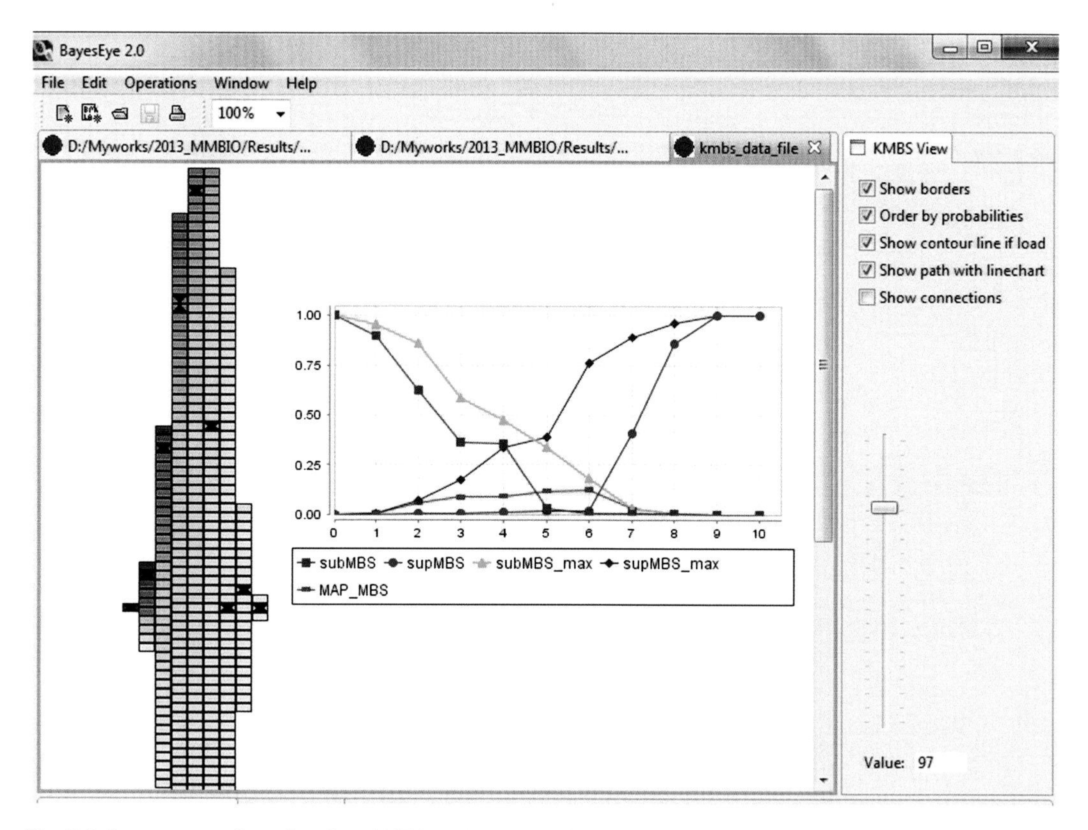

Fig. 2 Relevance map based on k-subMBS and k-supMBS subsets. The heatmap shows the possible subsets of various sizes. A column contains sets of the same size, the size increases from the *leftmost* column to the *right*. Cells are *colored* according to the probability of the represented set (from 0: *white* to 1: *red*). The graph details the probabilities of k-subMBS and k-supMBS related to a chosen path (cells marked with *black*) and also the subsets with the highest probability of a certain size (Color figure online)

leftmost node represents the empty set when there is no relevant predictor, and the rightmost node represents the complete set containing all the possible predictors when all of them are jointly relevant. An ordering of variables defines a path through this map by iteratively selecting edges between nodes according to the ordering. This path leads from the node representing the empty set to node representing the complete set. Along any path from the empty set to the complete set, sup-relevance is monotonically increasing from 0 to 1, and sub-relevance is decreasing from 1 to 0.

32. *Relevance tree*: This tool allows the analysis of the most relevant variable subsets. Figure 3 illustrates a relevance tree of the RA case study.
33. *Interaction*: Although k-subMBS posteriors allow the identification of stable patterns (of variables) having an arbitrary size *k*, they do not explicitly quantify whether a variable is part of a

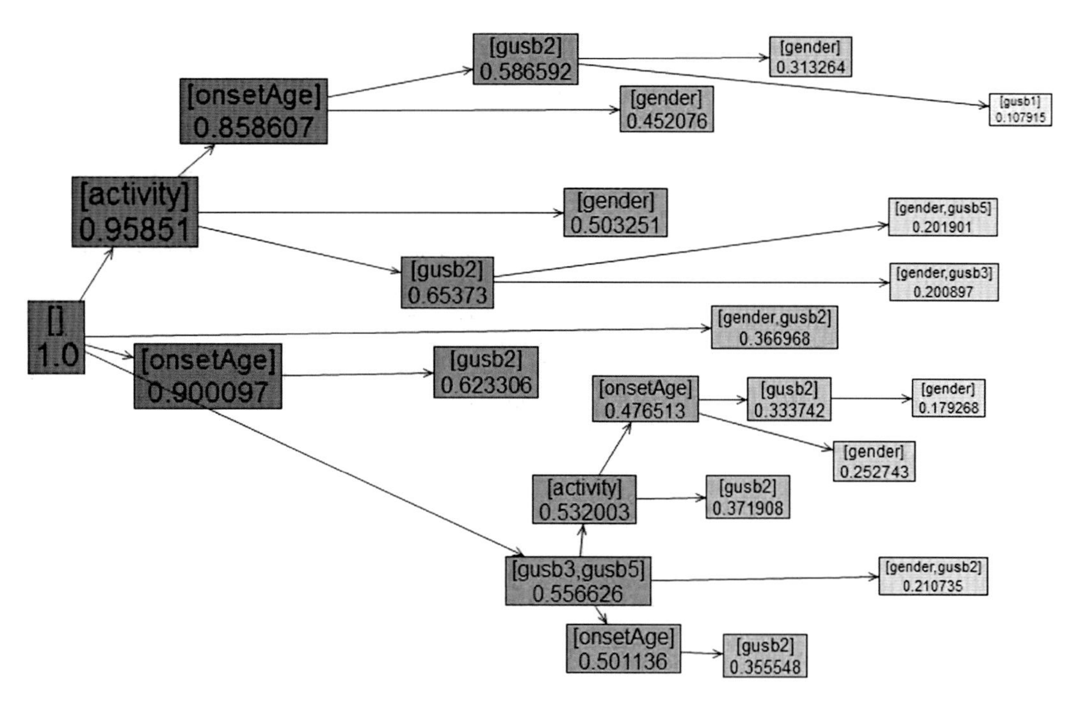

Fig. 3 Relevance tree. The subset members are indicated incrementally, and subset relations are in accordance with the edges (they are directed towards a larger subset containing the previous one). The size, color, and horizontal position of a set corresponds to its sub-relevance

given subset due to an interaction, or because it is relevant independently of others. Based on the decomposability of the posterior of the strong relevance of a set of variables such a score measuring interaction and redundancy can be constructed [15]. The interaction-redundancy score relies on the exact k-MBS posterior of a set of variables and its approximation as the product of the MBM probabilities of each member variable X_i in the given k-MBS, as if their occurrences were independent. Note that this model-level approach to interaction and redundancy formalizes the intuition that relevant input variables with decomposable roles at the parametric level appear independently in the model.

If the k-subMBS posterior of a set is larger than its approximation-based on MBM posteriors then this may indicate that the variables in the set have a joint parameterization expressing non-linear joint effects. In the opposite case, when the k-subMBS posterior is smaller than its approximation based on MBM posteriors, then this may indicate variable redundancy. This means that the joint presence of the redundant variables in the model is suppressed. Note that the interaction-redundancy scores corresponding to a given target do not appear to be related to the genetic linkage between the SNPs. Figure 4 illustrates the interaction and redundancy scores of

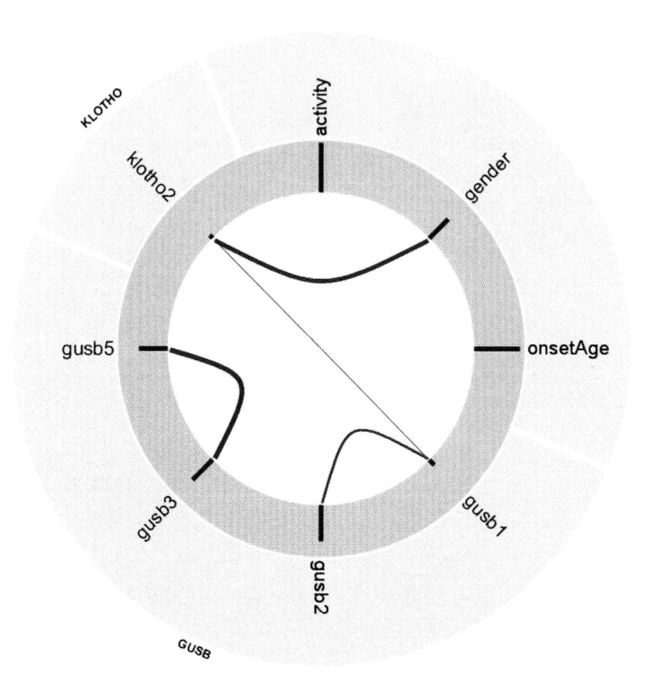

Fig. 4 Interaction and redundancy score visualization. The strong relevance of factors is indicated by bars in the *inner circle*, and the thickness of the edges is proportional with the strength of interactions (*red*) and redundancy (*blue*) (Color figure online)

variables of the RA case study. Results indicate that among relevant variables there is an interaction between SNPs GUSB3 and GUSB5.

34. *MBG visualization*: As in case of MBSs, the BayesEye software provides a tool for averaging and investigating MBGs. In case of large MBGs, a filter can be created using a Boolean query, which allows focusing on various parts of the MBG. Figure 5 shows an averaged MBG of the RA case study, which indicates that GUSB3 forms an interaction with GUSB5 and as such it is only relevant with respect to RA in conjunction with GUSB5.

35. *Bayesian network-based structure conditional effect size*: Effect size estimation is often preceded by structural exploration of associations. The main aim of GWAS studies is such a structural exploration. Since they use simple phenotype descriptors, they are not well suited for effect size estimation. CGAS studies on the other hand, although being generally positively biased, are more proper for effect size investigation as they use complex phenotypes and thus allow the conditioning on selected descriptors [74].

    The Bayesian framework allows both the structural exploration and the effect size computation. The estimation of odds ratios (OR) is a "full Bayesian" approach, as the structure level

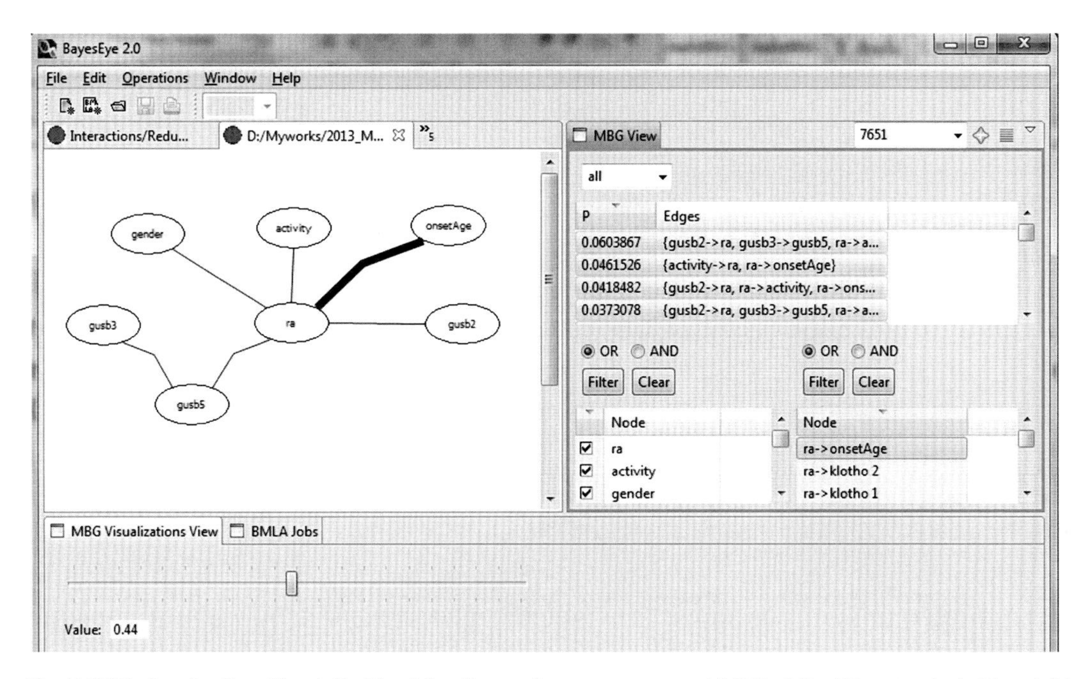

Fig. 5 MBG visualization. The *left side* of the figure shows an averaged MBG of the RA case study. The *right side* shows the filtering options and the list of individual MBGs

uncertainty is combined with the parameter level uncertainty using the Bayesian model averaging [32, 75] as follows

$$p(OR \mid D) = \sum_{G} p(OR \mid G, D) P(G \mid D),$$

in which $p(\mathrm{OR}|G,D)$ is the posterior distribution of the odds ratio given data D and a possible structure G, whereas $p(G|D)$ is the probability of that structure given data D. The Bayesian odds ratio $p(\mathrm{OR}|D)$ is a weighted average of posterior distributions $p(\mathrm{OR}|G,D)$.

The Bayesian framework allows the formulation of various effect size measures, such as pairwise measures and model-based approaches. A Bayesian pairwise approach can be based on the mean parameterization of Dirichlet prior with respect to the data [76]. The model-based approach using Bayesian networks allows a wide variety of effect strength concepts, such as concepts conditioned on structural properties, e.g., conditioned on Markov blankets.

36. *MBG-based odds ratio*: The posterior distribution $p(\mathrm{OR}|D)$ can be estimated by using Markov blanket graphs [76]. In contrast with a more complex graph structure, an MBG does not contain all the dependency relationships of variables, just those, in which the target is involved. This is a useful property, because generally the aim of effect size measures is to measure the effect of a factor on a target Y.

The Bayesian structure-based odds ratio $BS\text{-}OR(X, \Upsilon, \boldsymbol{MBG_\Upsilon})$ is such a measure, that uses MBGs to estimate the effect size of variables X_i with respect to a specific target Υ. An odds ratio is estimated for each subgraph $MBG_\Upsilon{}^j$ and weighted by the corresponding posterior of $MBG_\Upsilon{}^j$ given data D.

$$p(OR \mid D) \sim \sum_{MBG} p(OR \mid MBG, D) P(MBG \mid D),$$

where it is a crucial information, whether X_i is among the members of an $MBG_\Upsilon{}^i$. If X_i is not a member of an MBG, it means that in that particular model X_i is not relevant with respect to the target, in other words it has no relevant effect. This can be handled in multiple ways leading to various implementations. Averaging MBG conditional odds ratio takes such a non-relevant effect into account as a neutral odds ratio of 1 (for that MBG), whereas the selective MBG conditional odds ratio excludes that particular MBG from the averaging of effect sizes for variable X_i [76].

Figure 6 shows the posterior distributions of the Bayesian MBG-based odds ratios of SNPs GUSB3 and GUSB5, assuming a dominant genetic model. Both distributions are bimodal which suggests there are subpopulations where different mechanisms are dominant, e.g., in the case of GUSB3 the first peak is at the neutral odds ratio of 1 indicating negligible effect on RA, whereas the second peak at 1.2 indicates a subpopulation, in which the presence of the GUSB3 mutant allele presents a risk compared to the wild homozygote case. Note that a conventional odds ratio would detect a non-significant effect with a large confidence interval, without the details provided by the Bayesian odds ratio.

37. *Gene prioritization*: The systems-based analysis of relevance including effect strength estimation typically occurs in follow-up

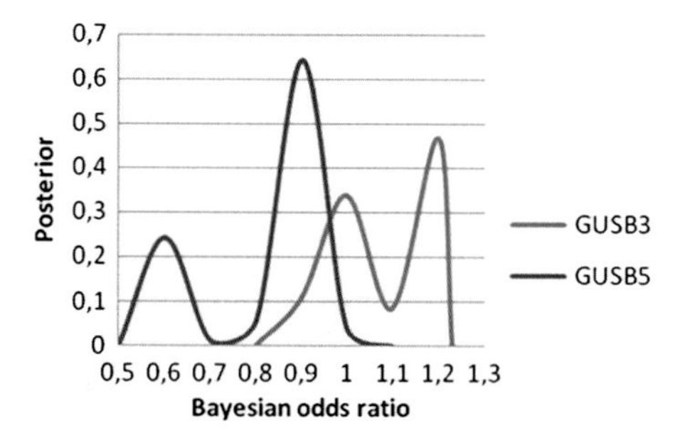

Fig. 6 Posterior distribution of Bayesian odds ratios. Each curve depicts the outline of a histogram of possible odds ratio values within the 95 % credible interval. Bayesian odds ratio values are shown on the horizontal axis, whereas related posterior probability values are displayed on the vertical axis

CGAS studies. In this case, the reintegration of the results to omic levels, e.g., understanding the functional aspects of associated, relevant predictors, is a serious challenge, which is called the "interpretational" or translational bottleneck. Thus, it is essential to support the extension of the focused scope of the CGAS to omic levels by incorporating large-scale, voluminous and heterogeneous data and knowledge. A wide range of technologies have been developed for the unification or fusion of fragmented biomedical knowledge [31], such as database integration [77], ontologies [78], natural language processing techniques [79], text-mining methods [80], network-based methods [81, 82], and multiple similarity-based methods [83, 84]. Computational gene prioritization (GP) became an encompassing term for this data and knowledge fusion problem, which can also be used for the omic extension of the CGAS results (for general guides for the application of GP, *see* for example [36, 85, 86]). We provide a step-by-step guide for the specialized application of GP methodology for interpreting genetic association studies, specifically for the aggregation of results to the gene level and for the derivation of disease specific kernels for GP (for tissue/disease specific GP, *see* refs. 87, 88). To access the combined BMLA-GP tool, please visit http://bioinformatics.mit.bme.hu/tools/GP.

38. *Omic information sources*: Kernel-based gene prioritization methods require background information to be expressed in the form of pairwise similarities between the genes. The special matrices that contain these similarity values and satisfy certain mathematical properties are called kernels. The information sources used in kernel construction can originate from a wide range of fields, for example text-mining of biomedical literature: functional annotations, pathways, or ontologies (e.g., Gene Ontology [78], Kyoto Encyclopedia of Genes and Genomes, KEGG [89]), PPIs (e.g., STRING [90], HPRD [91]), gene expression information (e.g., Gene Expression Omnibus, GEO[92]).

39. *STRING database*: In other words, one can view the information in the STRING database as an undirected weighted graph G, which consists of a set of vertices V (the genes corresponding to the proteins), and the set of weighted edges E (the combined score of the corresponding PPIs). The adjacency matrix of G is an $n \times n$ real matrix W (where n is the number of different genes), with W_{ij}= combined score of the interaction, if proteins i and j are connected, and 0 otherwise. If D is an $n \times n$ diagonal matrix with $D_{ii} = \sum_j W_{ij}$, then the Laplacian of G is $L = D - W$, and the Normalized Laplacian is $\tilde{L} = D^{-\frac{1}{2}} L D^{-\frac{1}{2}}$.

40. *Compound kernels*: More complex kernels can be constructed by incorporating other information sources, for example tissue specific gene expression profiles into the kernels (for a similar

concept in the case of network-based gene prioritization, *see* [87]).

41. *Query construction*: A crucial step in the kernel-based gene prioritization is query construction [36]. The comparison of the results of GPs using carefully crafted queries can highlight the content of the CGAS data. The a priori relevant genes can be identified using knowledge bases, such as OMIM [93], HugeNavigator [94], or the Catalog of Published Genome-Wide Association Studies [95]. Queries consisting of 5–30 elements have been shown to work well in typical use cases [36]. Some degree of heterogeneity is desirable, however, a high degree of heterogeneity in a query may lead to anomalies [88, 96].

42. *Query genes in the RA case study*: We constructed the GP query as a combination of a priori disease genes and the most relevant genes according to the CGAS results. First, we selected those genes that were associated with RA in more than ten publications according to the HugeNavigator knowledge base. Then, we added those genes to the query which were strongly relevant to RA based on the CGAS we performed. We used the following query genes in the analysis: TNF, PTPN22, IL1B, MTHFR, STAT4, IL10, IL1RN, TRAF1, IL6, TNFRSF1B, C5, MBL2, TNFAIP3, ERAP1, TNFRSF1A, HLA-A, IL2, IRF5, CCR5, CD40, HLA-C, IL4, TGFB1, CCL21, FCGR2A, LTA, ATIC, CIITA, IL18, VDR, ABCB1, IL1R1, MMP3, TAP2, TYMS, IFNG, IL2RA, IL4R, KIR3DL1, PDCD1, PRKCQ, GSTT1, MMP1, SLC19A1, SLC22A4, GUSB, KL.

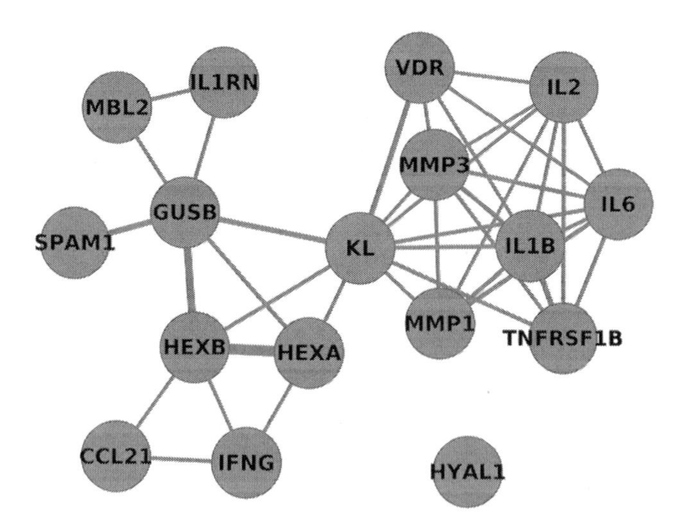

Fig. 7 Similarity network of selected top ranked genes of the RA case study. *Nodes* represent genes and *edges* represent similarities between the corresponding genes according to the kernels used in the analyses. The width of the *edges* are proportional to the pairwise similarity values

43. *Network analysis*: Figure 7 shows the similarity network of selected top ranked genes of the RA case study.
44. *Enrichment analysis*: Gene sets can originate from for example the Molecular Signature Database (MSigDB) [97].
45. *Semantic publishing*: The last phases in the GAS workflow are the dissemination and application of GAS results. In a narrow sense, translational research covers the phases from interpretation to evaluation, to questions of repository and dissemination, and to real-world applications. Because of the accumulation of the results, their storage and dissemination is an increasingly important issue, especially to cope with biases, e.g., the positive publication bias, and to support meta-analysis and automated knowledge integration [98–109]. Besides recommendations for reporting GAS results, semantic publishing and structured digital abstracts were proposed to bridge the gap between free-text publications and databases [110–117]. Despite the potential advantage of this systematic approach, there are no standardized solutions yet, which could change rapidly, because of the need from evidence-based personalized medicine [118, 119]. The Bayesian application of PGMs in relevance analysis, e.g., the BMLA approach for complex phenotypes, fits this trend, as their results can be interpreted as probabilistic, semantically transparent statements, i.e., forming a probabilistic knowledge base, which could be directly used in complex queries and even in the fusion of multiple analyses, i.e., meta-analysis [18].

References

1. Dermitzakis E (2008) From gene expression to disease risk. Nat Genet 40:492–493
2. Maher B (2008) Personal genomes: the case of the missing heritability. Nature 456:18–21
3. Joober R (2011) The 1000 Genomes Project: deep genomic sequencing waiting for deep psychiatric phenotyping. J Psychiatry Neurosci 36:147–149
4. Moreau Y, Antal P, Fannes G, De Moor B (2003) Probabilistic graphical models for computational biomedicine. Methods Inf Med 42:161–168
5. Saeys Y, Inza I, Larranaga P (2007) A review of feature selection techniques in bioinformatics. Bioinformatics 23:2507–2517
6. Rodin A, Boerwinkle E (2005) Mining genetic epidemiology data with Bayesian networks I: Bayesian networks and example application (plasma apoE levels). Bioinformatics 21:3273–3278
7. Verzilli C, Stallard N, Whittaker J (2006) Bayesian graphical models for genomewide association studies. Am J Hum Genet 79: 100–112
8. Mourad R, Sinoquet C, Leray P (2012) Probabilistic graphical models for genetic association studies. Brief Bioinform 13: 20–33
9. Li W, Wang M, Irigoyen P, Gregersen P (2006) Inferring causal relationships among intermediate phenotypes and biomarkers: a case study of rheumatoid arthritis. Bioinformatics 22:1503–1507
10. Xing H, McDonagh P, Bienkowska J, Cashorali T, Runge K, Miller R, DeCaprio D, Church B, Roubenoff R, Khalil I, Carulli J (2011) Causal modeling using network ensemble simulations of genetic and gene expression data predicts genes involved in rheumatoid arthritis. PLoS Comput Biol 7:e1001105
11. Han B, Park M, Chen X (2010) A Markov blanket-based method for detecting causal SNPs in GWAS. BMC Bioinformatics 11 Suppl 3:S5
12. Jiang X, Barmada MM, Visweswaran S (2010) Identifying genetic interactions in genome-wide data using Bayesian networks. Genet Epidemiol 34:575–581
13. Fridley B (2009) Bayesian variable and model selection methods for genetic association studies. Genet Epidemiol 33:27–37

14. Antal P, Millinghoffer A, Hullám G, Hajós G, Sárközy P, Szalai C, Falus A (in press) Bayesian, systems-based, multilevel analysis of biomarkers of complex phenotypes: from interpretation to decisions. In: Sinoquet C, Mourad R (eds) Probabilistic graphical models for genetics, genomics and postgenomics. ISBN: 978-0-19-870902-2, Oxford University Press
15. Antal P, Millinghoffer A, Hullam G, Szalai C, Falus A (2008) A Bayesian view of challenges in feature selection: feature aggregation, multiple targets, redundancy and interaction. In: Saeys Y, Liu H, Inza I, Wehenkel L, Van de Peer Y (eds) New challenges for feature selection in data mining and knowledge discovery (FSDM), JMLR workshop and conference proceedings, September 15, 2008, Antwerp, Belgium, pp 74–89
16. Antal P, Hullám G, Gézsi A, Millinghoffer A (2006) Learning complex Bayesian network features for classification. In: Third European workshop on probabilistic graphical model, Prague, pp 9–16
17. Pal Z, Antal P, Millinghoffer A, Hullam G, Paloczi K, Toth S, Gabius H, Molnar M, Falus A, Buzas E (2010) A novel galectin-1 and interleukin 2 receptor beta haplotype is associated with autoimmune myasthenia gravis. J Neuroimmunol 229:107–111
18. Sarkozy P, Marx P, Millinghoffer A, Varga G, Szekely A, Nemoda Z, Demetrovics Z, Sasvari-Szekely M, Antal P (2011) Bayesian data analytic knowledge bases for genetic association studies. In: Arjen Hommersom PL (ed) The 13th conference on artificial intelligence in medicine (AIME'11): probabilistic problem solving in biomedicine, July 2–6, 2011, Bled, Slovenia, pp 55–66
19. Lautner-Csorba O, Gezsi A, Semsei AF, Antal P, Erdelyi DJ, Schermann G, Kutszegi N, Csordas K, Hegyi M, Kovacs G, Falus A, Szalai C (2012) Candidate gene association study in pediatric acute lymphoblastic leukemia evaluated by Bayesian network based Bayesian multilevel analysis of relevance. BMC Med Genomics 5:42
20. Ungvari I, Hullam G, Antal P, Kiszel P, Gezsi A, Hadadi E, Virag V, Hajos G, Millinghoffer A, Nagy A, Kiss A, Semsei A, Temesi G, Melegh B, Kisfali P, Szell M, Bikov A, Galffy G, Tamasi L, Falus A, Szalai C (2012) Evaluation of a partial genome screening of two asthma susceptibility regions using Bayesian network based Bayesian multilevel analysis of relevance. PLoS One 7:e33573
21. Varga G, Szekely A, Antal P, Sarkozy P, Nemoda Z, Demetrovics Z, Sasvari-Szekely M (2012) Additive effects of serotonergic and dopaminergic polymorphisms on trait impulsivity. Am J Med Genet B Neuropsychiatr Genet 159B(3):281–288
22. Lautner-Csorba O, Gézsi A, Erdélyi D, Hullám G, Antal P, Semsei Á, Kutszegi N, Kovács G, Falus A, Szalai C (2013) Roles of genetic polymorphisms in the folate pathway in childhood acute lymphoblastic leukemia evaluated by Bayesian relevance and effect size analysis. PLoS One 8:e69843
23. Vereczkei A, Demetrovics Z, Szekely A, Sarkozy P, Antal P, Szilagyi A, Sasvari-Szekely M, Barta C (2013) Multivariate analysis of dopaminergic gene variants as risk factors of heroin dependence. PLoS One 8:e66592
24. Stephens M, Balding D (2009) Bayesian statistical methods for genetic association studies. Nat Rev Genet 10:681–690
25. Beaumont M, Rannala B (2004) The Bayesian revolution in genetics. Nat Rev Genet 5:251–261
26. Roeder K, Devlin B, Wasserman L (2007) Improving power in genome-wide association studies: weights tip the scale. Genet Epidemiol 31:741–747
27. Curtis D, Vine A, Knight J (2007) A pragmatic suggestion for dealing with results for candidate genes obtained from genome wide association studies. BMC Genet 8:20
28. Jiang X, Barmada M, Cooper G, Becich M (2011) A Bayesian method for evaluating and discovering disease loci associations. PLoS One 6:e22075
29. Saccone S, Saccone N, Swan G, Madden P, Goate A, Rice J, Bierut L (2008) Systematic biological prioritization after a genome-wide association study: an application to nicotine dependence. Bioinformatics 24: 1805–1811
30. Saccone S, Bolze R, Thomas P, Quan J, Mehta G, Deelman E, Tischfield J, Rice J (2010) SPOT: a web-based tool for using biological databases to prioritize SNPs after a genome-wide association study. Nucleic Acids Res 38:W201–W209
31. Saccone S, Chesler E, Haendel M (2012) Applying in silico integrative genomics to genetic studies of human disease. Bioinformatics of Behavior: Part 1 103: 133–156
32. Madigan D, Andersson S, Perlman M, Volinsky C (1996) Bayesian model averaging and model selection for Markov equivalence classes of acyclic digraphs. Comm Stat Theor Methods 25:2493–2519
33. Friedman N, Koller D (2003) Being Bayesian about network structure. A Bayesian approach to structure discovery in Bayesian networks. Mach Learn 50:95–125
34. Efron B (2013) Bayes' theorem in the 21st century. Science 340:1177–1178
35. Pearl J (2000) Causality: models, reasoning, and inference. Cambridge University Press, New York

36. Moreau Y, Tranchevent L (2012) Computational tools for prioritizing candidate genes: boosting disease gene discovery. Nat Rev Genet 13:523–536
37. Pettersson F, Anderson C, Clarke G, Barrett J, Cardon L, Morris A, Zondervan K (2009) Marker selection for genetic case-control association studies. Nat Protoc 4:743–752
38. Nsengimana J, Bishop DT (2012) Design considerations for genetic linkage and association studies. Methods Mol Biol 850:237–262
39. Friedman N, Yakhini Z (1996) On the sample complexity of learning Bayesian networks. In: Horvitz E, Jensen F (eds.) UAI'96: Proceedings of the Twelfth international conference on Uncertainty in artificial intelligence, August 1-4, 1996, Portland, Oregon, USA, pp 274–282
40. Hullám G, Antal P, Millinghoffer A, Szalai C, Falus A (2010) Evaluation of a Bayesian model-based approach in GA studies. In: JMLR workshop and conference proceeding, pp 30–43
41. Vittinghoff E, McCulloch C (2007) Relaxing the rule of ten events per variable in logistic and Cox regression. Am J Epidemiol 165:710–718
42. Stahl EA, Wegmann D, Trynka G, Gutierrez-Achury J, Do R, Voight BF, Kraft P, Chen R, Kallberg HJ, Kurreeman FA, Diabetes Genetics Replication and Meta-analysis Consortium, Myocardial Infarction Genetics Consortium (2012) Bayesian inference analyses of the polygenic architecture of rheumatoid arthritis. Nat Genet 44:483–489
43. Skapenko A, Prots I, Schulze-Koops H (2009) Prognostic factors in rheumatoid arthritis in the era of biologic agents. Nat Rev Rheumatol 5:491–496
44. Ortutay Z, Polgar A, Gomor B, Geher P, Lakatos T, Glant T, Gay R, Gay S, Pallinger E, Farkas C, Farkas E, Tothfalusi L, Kocsis K, Falus A, Buzas E (2003) Synovial fluid exoglycosidases are predictors of rheumatoid arthritis and are effective in cartilage glycosaminoglycan depletion. Arthritis Rheum 48:2163–2172
45. Pasztoi M, Nagy G, Geher P, Lakatos T, Toth K, Wellinger K, Pocza P, Gyorgy B, Holub M, Kittel A, Paloczy K, Mazan M, Nyirkos P, Falus A, Buzas E (2009) Gene expression and activity of cartilage degrading glycosidases in human rheumatoid arthritis and osteoarthritis synovial fibroblasts. Arthritis Res Ther 11:R68
46. Wigginton J, Cutler D, Abecasis G (2005) A note on exact tests of Hardy-Weinberg equilibrium. Am J Hum Genet 76:887–893
47. Marchini J, Howie B, Myers S, McVean G, Donnelly P (2007) A new multipoint method for genome-wide association studies by imputation of genotypes. Nat Genet 39:906–913
48. Dempster A, Laird N, Rubin D (1977) Maximum likelihood from incomplete data via EM algorithm. J Roy Stat Soc B Stat Methods 39:1–38
49. Tanner M, Wong W (2010) From EM to data augmentation: the emergence of MCMC Bayesian computation in the 1980s. Stat Sci 25:506–516
50. Gelman A (1995) Bayesian data analysis, 1st edn. Chapman & Hall, New York
51. Barrett J, Fry B, Maller J, Daly M (2005) Haploview: analysis and visualization of LD and haplotype maps. Bioinformatics 21:263–265
52. Zhang J, Rowe W, Struewing J, Buetow K (2002) HapScope: a software system for automated and visual analysis of functionally annotated haplotypes. Nucleic Acids Res 30:5213–5221
53. Gu S, Pakstis A, Kidd K (2005) HAPLOT: a graphical comparison of haplotype blocks, tagSNP sets and SNP variation for multiple populations. Bioinformatics 21:3938–3939
54. Davidovich O, Kimmel G, Shamir R (2007) GEVALT: an integrated software tool for genotype analysis. BMC Bioinformatics 8:36
55. Stephens M, Smith N, Donnelly P (2001) A new statistical method for haplotype reconstruction from population data. Am J Hum Genet 68:978–989
56. Mourad R, Sinoquet C, Leray P (2011) A hierarchical Bayesian network approach for linkage disequilibrium modeling and data-dimensionality reduction prior to genome-wide association studies. BMC Bioinformatics 12:16
57. Kost J, McDermott M (2002) Combining dependent P-values. Stat Probab Lett 60: 183–190
58. Zhang F, Guo X, Wu S, Han J, Liu Y, Shen H, Deng H (2012) Genome-wide pathway association studies of multiple correlated quantitative phenotypes using principle component analyses. PLoS One 7:e53320
59. Friedman N, Goldszmidt M (1996) Discretizing continuous attributes while learning Bayesian networks. In: Saitta L (ed) Thirteenth international conference on machine learning, (ICML '96). Morgan Kaufmann, Bari, pp 157–165
60. Hullam G, Antal P (2013) The effect of parameter priors on Bayesian relevance and effect size measures. Periodica Polytechnica Electrical Engineering and Computer Science 57:35–48
61. Cooper G, Herskovits E (1992) A Bayesian method for the induction of probabilistic networks from data. Mach Learn 9:309–347
62. Silander T, Kontkanen P, Myllymaki P (2007) On sensitivity of the MAP Bayesian network structure to the equivalent sample size parameter. AUAI Press, Corvallis, OR, pp 360–367
63. Ueno M (2010) Learning networks determined by the ratio of prior and data. AUAI Press, Corvallis, OR, pp 598–605

64. Bouckaert RR (1994) Properties of Bayesian belief network learning algorithms. Morgan Kaufmann, San Francisco, CA, pp 102–109
65. Buntine WL (1991) Theory refinement on Bayesian networks. In: D'Ambrosio B, Smets P (eds.): UAI '91: Proceedings of the Seventh Annual Conference on Uncertainty in Artificial Intelligence, July 13-15, 1991, UCLA, Los Angeles, CA, USA, pp 52–60
66. Heckerman D, Geiger D, Chickering D (1995) Learning Bayesian networks—the combination of knowledge and statistical-data. Mach Learn 20:197–243
67. Giudici P, Castelo R (2003) Improving Markov Chain Monte Carlo model search for data mining. Mach Learn 50:127–158
68. Chen M-H, Shao Q-M, Ibrahim JG (2000) Monte Carlo methods in Bayesian computation. Springer, New York
69. Pearl J (1988) Probabilistic reasoning in intelligent systems: networks of plausible inference. Morgan Kaufmann Publishers, San Mateo, CA
70. Kohavi R, John G (1997) Wrappers for feature subset selection. Artif Intell 97:273–324
71. Tsamardinos I, Aliferis C (2003) Towards Principled Feature Selection: Relevancy, Filters, and Wrappers. In: Bishop CM, Frey BJ (eds.) Proc. of the Ninth International Workshop on Artificial Intelligence and Statistics, January 3-6, 2003, Morgan Kaufmann Publishers, Key West, FL, USA, pp 334–342
72. O'Hara R, Sillanpaa M (2009) A review of Bayesian variable selection methods: what, how and which. Bayesian Anal 4:85–117
73. Kooperberg C, Ruczinski I (2005) Identifying interacting SNPs using Monte Carlo logic regression. Genet Epidemiol 28:157–170
74. Ioannidis J (2008) Why most discovered true associations are inflated. Epidemiology 19: 640–648
75. Hoeting JA, Madigan D, Raftery AE, Volinsky CT (1999) Bayesian model averaging: a tutorial. Stat Sci 14:382–417
76. Hullam G, Antal P (2012) Estimation of effect size posterior using model averaging over Bayesian network structures and parameters. In: The sixth European workshop on probabilistic graphical models (PGM2012), Granada, Spain
77. Stein L (2003) Integrating biological databases. Nat Rev Genet 4:337–345
78. Ashburner M, Ball C, Blake J, Botstein D, Butler H, Cherry J, Davis A, Dolinski K, Dwight S, Eppig J, Harris M, Hill D, Issel-Tarver L, Kasarskis A, Lewis S, Matese J, Richardson J, Ringwald M, Rubin G, Sherlock G, The Gene Ontology Consortium (2000) Gene ontology: tool for the unification of biology. Nat Genet 25:25–29
79. Liekens A, De Knijf J, Daelemans W, Goethals B, De Rijk P, Del-Favero J (2011) BioGraph: unsupervised biomedical knowledge discovery via automated hypothesis generation. Genome Biol 12:R57
80. Glenisson P, Coessens B, Van Vooren S, Mathys J, Moreau Y, De Moor B (2004) TXTGate: profiling gene groups with text-based information. Genome Biol 5:R43
81. Kohler S, Bauer S, Horn D, Robinson P (2008) Walking the interactome for prioritization of candidate disease genes. Am J Hum Genet 82:949–958
82. Lee I, Blom U, Wang P, Shim J, Marcotte E (2011) Prioritizing candidate disease genes by network-based boosting of genome-wide association data. Genome Res 21:1109–1121
83. Lanckriet G, De Bie T, Cristianini N, Jordan M, Noble W (2004) A statistical framework for genomic data fusion. Bioinformatics 20: 2626–2635
84. De Bie T, Tranchevent L, Van Oeffelen L, Moreau Y (2007) Kernel-based data fusion for gene prioritization. Bioinformatics 23: I125–I132
85. Bromberg Y (2013) Chapter 15: disease gene prioritization. PLoS Comput Biol 9:e1002902
86. Doncheva N, Kacprowski T, Albrecht M (2012) Recent approaches to the prioritization of candidate disease genes. Wiley Interdiscip Rev Syst Biol Med 4:429–442
87. Magger O, Waldman Y, Ruppin E, Sharan R (2012) Enhancing the prioritization of disease-causing genes through tissue specific protein interaction networks. PLoS Comput Biol 8:e1002690
88. Navlakha S, Kingsford C (2010) The power of protein interaction networks for associating genes with diseases. Bioinformatics 26: 1057–1063
89. Kanehisa M, Goto S (2000) KEGG: Kyoto encyclopedia of genes and genomes. Nucleic Acids Res 28:27–30
90. Franceschini A, Szklarczyk D, Frankild S, Kuhn M, Simonovic M, Roth A, Lin J, Minguez P, Bork P, von Mering C, Jensen L (2013) STRING v9.1: protein-protein interaction networks, with increased coverage and integration. Nucleic Acids Res 41:D808–D815
91. Prasad T, Goel R, Kandasamy K, Keerthikumar S, Kumar S, Mathivanan S, Telikicherla D, Raju R, Shafreen B, Venugopal A, Balakrishnan L, Marimuthu A, Banerjee S, Somanathan D, Sebastian A, Rani S, Ray S, Kishore C, Kanth S, Ahmed M, Kashyap M, Mohmood R, Ramachandra Y, Krishna V, Rahiman B, Mohan S, Ranganathan P, Ramabadran S, Chaerkady R, Pandey A (2009) Human protein reference database-2009 update. Nucleic Acids Res 37:D767–D772
92. Edgar R, Domrachev M, Lash A (2002) Gene expression omnibus: NCBI gene expression

and hybridization array data repository. Nucleic Acids Res 30:207–210
93. McKusick-Nathans Institute for Genetic Medicine. Online Mendelian Inheritance in Man, OMIM®. Johns Hopkins University, Baltimore, MD. http://omim.org/
94. Yu W, Gwinn M, Clyne M, Yesupriya A, Khoury M (2008) A navigator for human genome epidemiology. Nat Genet 40: 124–125
95. Hindorff L, Sethupathy P, Junkins H, Ramos E, Mehta J, Collins F, Manolio T (2009) Potential etiologic and functional implications of genome-wide association loci for human diseases and traits. Proc Natl Acad Sci U S A 106:9362–9367
96. Arany A, Bolgar B, Balogh B, Antal P, Matyus P (2013) Multi-aspect candidates for repositioning: data fusion methods using heterogeneous information sources. Curr Med Chem 20:95–107
97. Subramanian A, Tamayo P, Mootha V, Mukherjee S, Ebert B, Gillette M, Paulovich A, Pomeroy S, Golub T, Lander E, Mesirov J (2005) Gene set enrichment analysis: a knowledge-based approach for interpreting genome-wide expression profiles. Proc Natl Acad Sci U S A 102:15545–15550
98. Attia J, Ioannidis J, Thakkinstian A, McEvoy M, Scott R, Minelli C, Thompson J, Infante-Rivard C, Guyatt G (2009) How to use an article about genetic association a: background concepts. JAMA 301:74–81
99. Attia J, Ioannidis J, Thakkinstian A, McEvoy M, Scott R, Minelli C, Thompson J, Infante-Rivard C, Guyatt G (2009) How to use an article about genetic association B: are the results of the study valid? JAMA 301:191–197
100. Attia J, Ioannidis J, Thakkinstian A, McEvoy M, Scott R, Minelli C, Thompson J, Infante-Rivard C, Guyatt G (2009) How to use an article about genetic association C: what are the results and will they help me in caring for my patients? JAMA 301:304–308
101. Huang J, Mirel D, Pugh E, Xing C, Robinson P, Pertsemlidis A, Ding L, Kozlitina J, Maher J, Rios J, Story M, Marthandan N, Scheuermann R (2011) Minimum information about a genotyping experiment (MIGEN). Stand Genomic Sci 5:224–229
102. Janssens A, Ioannidis J, van Duijn C, Little J, Khoury M, Grp G (2011) Strengthening the reporting of Genetic Risk Prediction Studies: the GRIPS statement. Genet Med 13:453–456
103. Little J, Higgins J, Ioannidis J, Moher D, Gagnon F, von Elm E, Khoury M, Cohen B, Davey-Smith G, Grimshaw J, Scheet P, Gwinn M, Williamson R, Zou G, Hutchings K, Johnson C, Tait V, Wiens M, Golding J, van Duijn C, McLaughlin J, Paterson A, Wells G, Fortier I, Freedman M, Zecevic M, King R, Infante-Rivard C, Stewart A, Birkett N (2009) STrengthening the REporting of Genetic Association studies (STREGA)—an extension of the STROBE statement. Eur J Clin Invest 39:247–266
104. Ioannidis J, Khoury M (2011) Improving validation practices in "Omics" research. Science 334:1230–1232
105. Colhoun H, McKeigue P, Smith G (2003) Problems of reporting genetic associations with complex outcomes. Lancet 361:865–872
106. Shi G, Boerwinkle E, Morrison A, Gu C, Chakravarti A, Rao D (2011) Mining gold dust under the genome wide significance level: a two-stage approach to analysis of GWAS. Genet Epidemiol 35:111–118
107. Province M, Borecki I (2007) Gathering the gold dust: identification small-effect complex trait genes. Genet Epidemiol 31:611–612
108. Evangelou E, Ioannidis J (2013) Meta-analysis methods for genome-wide association studies and beyond. Nat Rev Genet 14:379–389
109. Pers T, Hansen N, Lage K, Koefoed P, Dworzynski P, Miller M, Flint T, Mellerup E, Dam H, Andreassen O, Djurovic S, Melle I, Borglum A, Werge T, Purcell S, Ferreira M, Kouskoumvekaki I, Workman C, Hansen T, Mors O, Brunak S (2011) Meta-analysis of heterogeneous data sources for genome-scale identification of risk genes in complex phenotypes. Genet Epidemiol 35:318–332
110. Little J, Higgins J, Ioannidis J, Moher D, Gagnon F, von Elm E, Khoury M, Cohen B, Davey-Smith G, Grimshaw J, Scheet P, Gwinn M, Williamson R, Zou G, Hutchings K, Johnson C, Tait V, Wiens M, Golding J, van Duijn C, McLaughlin J, Paterson A, Wells G, Fortier I, Freedman M, Zecevic M, King R, Infante-Rivard C, Stewart A, Birkett N (2009) Strengthening the reporting of genetic association studies (STREGA): an extension of the STROBE Statement. Hum Genet 125:131–151
111. Shotton D (2009) Semantic publishing: the coming revolution in scientific journal publishing. Learn Publish 22:85–94
112. Shotton D, Portwin K, Klyne G, Miles A (2009) Adventures in semantic publishing: exemplar semantic enhancements of a research article. PLoS Comput Biol 5:e1000361
113. Seringhaus M, Gerstein M (2008) Manually structured digital abstracts: a scaffold for automatic text mining. FEBS Lett 582:1170
114. Gerstein M, Seringhaus M, Fields S (2007) Structured digital abstract makes text mining easy. Nature 447:142
115. Seringhaus M, Gerstein M (2007) Publishing perishing? Towards tomorrow's information architecture. BMC Bioinformatics 8:17

116. Bourne P (2005) Will a biological database be different from a biological journal? PLoS Comput Biol 1:179–181
117. Gerstein M (1999) E-publishing on the web: promises, pitfalls, and payoffs for bioinformatics. Bioinformatics 15:429–431
118. Goddard K, Knaus W, Whitlock E, Lyman G, Feigelson H, Schully S, Ramsey S, Tunis S, Freedman A, Khoury M, Veenstra D (2012) Building the evidence base for decision making in cancer genomic medicine using comparative effectiveness research. Genet Med 14:633–642
119. Gwinn M, Grossniklaus D, Yu W, Melillo S, Wulf A, Flome J, Dotson W, Khoury M (2011) Horizon scanning for new genomic tests. Genet Med 13:161–165

Chapter 15

Sample Processing, Protocol, and Statistical Analysis of the Time-of-Flight Secondary Ion Mass Spectrometry (ToF-SIMS) of Protein, Cell, and Tissue Samples

Goncalo Barreto, Antti Soininen, Tarvo Sillat, Yrjö T. Konttinen, and Emilia Kaivosoja

Abstract

Time-of-flight secondary ion mass spectrometry (ToF-SIMS) is increasingly being used in analysis of biological samples. For example, it has been applied to distinguish healthy and osteoarthritic human cartilage. This chapter discusses ToF-SIMS principle and instrumentation including the three modes of analysis in ToF-SIMS. ToF-SIMS sets certain requirements for the samples to be analyzed; for example, the samples have to be vacuum compatible. Accordingly, sample processing steps for different biological samples, i.e., proteins, cells, frozen and paraffin-embedded tissues and extracellular matrix for the ToF-SIMS are presented. Multivariate analysis of the ToF-SIMS data and the necessary data preprocessing steps (peak selection, data normalization, mean-centering, and scaling and transformation) are discussed in this chapter.

Key words Time-of-flight secondary ion mass spectrometry, Principal component analysis

1 Introduction

Time-of-flight secondary ion mass spectrometry (ToF-SIMS) provides spectroscopy for characterization of chemical composition, imaging for determining the distribution of chemical species, and depth profiling for thin film characterization. It was initially designed to evaluate the purity of materials in semiconductor and other industry, but has recently been increasingly applied for biological analysis. No other spectroscopic method has shown the versatility of ToF-SIMS, which allows the detection of mainly lipids and inorganic substances in biological samples with extremely high spatial resolution and sensitivity and high specificity. ToF-SIMS provides a molecular fingerprint of each surface and basic knowledge about the chemical composition of specific structures in cells and tissue. It enables the observation of disease-induced changes in local chemical composition and mapping of drug-induced

Shunichi Shiozawa (ed.), *Arthritis Research: Methods and Protocols*, Methods in Molecular Biology, vol. 1142, DOI 10.1007/978-1-4939-0404-4_15, © Springer Science+Business Media New York 2014

chemical changes. For example, the method has proven to be as a powerful tool to investigate disease-related tissue spatial lipid signatures [1]. Moreover, ToF-SIMS combined with multivariate analysis allows identification of a particular protein from a multicomponent protein film [2, 3] and enables identification of different types of cells and tissues. For example, discrimination between different yeast strains [4], different breast cancer cell types [5], bacterial and eukaryotic cells [6], different tissue types from mouse embryos [7], and decellularized small intestine and decellularized esophagus [8] has been achieved. This kind of identification of cell/tissue types could be used to improve disease diagnosis and aid in evaluation of the prognosis. For example, ToF-SIMS together with multivariate analysis has been applied to distinguishing healthy and osteoarthritic (OA) human cartilage [9]. The spatial distribution of cholesterol-related peaks exhibited a remarkable difference between healthy and OA cartilages and an accumulation of calcium and phosphate ions exclusively in areas surrounding the chondrocyte in OA tissues was observed. ToF-SIMS has also been used as a novel technique to identify the differentiation stages of hematopoietic cells [10]. This was possible by comparing the specific amino acid to lipid ratio signatures of B cells, common lymphoid progenitors (CLPs), and hematopoietic stem and progenitor cells (HSPCs). The same study showed that the B cells, CLPs, and HSPCs isolated from old mice had lower amino acid to lipid ratios when compared to cells isolated from young mice.

1.1 ToF-SIMS Principle and Instrumentation

Secondary Ion Mass Spectrometry (SIMS) is a solid surface analysis technique with a high surface sensitivity for the determination of surface composition, contaminant analysis and for depth profiling of the uppermost atomic surface layers of a sample. Of the surface analysis techniques, SIMS is the only one sensitive to all elements. SIMS is used to define the elemental and molecular composition of solid materials, and it also provides methods of visualizing the 2D and 3D composition of solids. SIMS technique uses an energetic ion beam to sputter the atomic and molecular constituents from a surface in a very controlled manner. These constituents include atoms, molecules, and molecular fragments that are characteristic of the surface composition within each volume element sputtered by the ion beam used. The sputtering ions are referred to as the primary ions while the ions produced by sputtering the solid are referred as the secondary ions. SIMS analyses are divided into two broad categories known as dynamic and static. In the dynamic mode, a relatively intense primary ion beam sputters the sample surface at sputter rates ranging from 1 to 20 atom layers/s. This high sputter rate provides a very useful method for determining the depth concentration of different elements in a solid and is the most common method of SIMS analysis. Since the primary ion beam is eroding away the surface at a high rate, most molecular or chemical bonding information is rapidly destroyed, and hence the most common secondary ions detected (in a dynamic SIMS analysis) are elemental ions or clusters of elemental ions. Static mode SIMS utilizes a very low intensity primary ion beam, and static mode analyses are typically completed before a single monolayer has been removed from the surface. Most static analyses are stopped (or used in pulsed mode) before the top surface layer has been chemically damaged or eroded; under these conditions, molecular and molecular fragment ions characteristic of the chemical structure of the surface are often detected. Thus, static SIMS is best suited for near-surface analysis of molecular composition or chemical structure information, while dynamic secondary ion mass spectrometry provides the best technique for bulk and depth profile elemental analysis [11–14].

Time-of-Flight Secondary Ion Mass Spectrometry (ToF-SIMS) uses a pulsed primary ion beam to detach elemental and molecular species from a sample surface. The secondary ions are accelerated into a mass spectrometer, where they are time-of-flight mass analyzed. ToF mass spectrometry is based on the fact that ions with the same energy but different masses travel with different velocities [15]. ToF-SIMS has the smallest sampling depth for the ultimate in surface and depth specificity when investigating the microscopic detail of the 2D and 3D chemistry of solids. This is important for the successful analysis of small traces of contaminations and also very thin nanofilms or multilayers. ToF-SIMS system normally consists of the following components: an ultrahigh vacuum system, a primary ion gun (with desired ion source), the flight path, and the mass detector system, and also ToF-SIMS instrument is normally equipped with a powerful computer and software for system control and analysis. There are three different modes of analysis in ToF-SIMS:

1. ToF-SIMS spectroscopy: The primary ion impacts the sample surface. The emitted secondary ions are extracted into the ToF analyzer by applying a high voltage potential between the sample surface and the mass analyzer. ToF-SIMS spectra are generated using a pulsed primary ion source. Secondary ions travel through the ToF analyzer with different velocities depending on their mass-to-charge ratio. For each primary ion pulse, full mass spectra are acquired to determine the elemental and molecular species on a surface.
2. ToF-SIMS ion imaging: Chemical images are generated by collecting a mass spectrum at different positions as the primary ion beam is moved across the sample surface area. Images are acquired to visualize the distribution of individual species on the surface. With ToF-SIMS imaging mass spectrometry, it is possible not only to screen many chemical substances to find those that change in disease, but also to study the detailed spatial and temporal relationships among the substances

3. ToF-SIMS depth profiling: ToF-SIMS is capable of shallow sputter depth profiling. An ion gun is operated in the DC mode during the sputtering phase in order to remove material, and the same ion gun or a second ion gun is operated in the pulsed mode for acquisition phase. Depth profiling by ToF-SIMS allows monitoring of all species of interest simultaneously with high mass resolution. These depth profiles are used to determine the distribution of different chemical species as a function of depth from the surface.

1.2 ToF-SIMS Analysis

ToF-SIMS spectroscopy from biological samples, e.g., protein films, tissues, cell "footprints," results usually in quite complex spectra with large number of different peaks. Multivariate analysis with principal component analysis (PCA) has been most commonly used in interpretation of such data [16, 17], although also more computationally demanding maximum autocorrelation factor method has been suggested for biomaterial analysis [18]. With PCA the data of possibly correlated variables is transformed to a new set of (and possible fewer) uncorrelated variables (so-called principal components) [19, 20]. The first principle component accounts for the largest amount of variation in data and each following principal component for successfully lesser amount of variation. Therefore, even the first few principal components may already show the hidden trends or grouping inside the data set. For protein analysis the usage of only limited set of different amino acid secondary ion fragment peaks has been suggested [15, 16, 21, 22]. This way the variation in peaks caused from inorganic components from buffers, impurities, contamination, etc. as well as from some oxygenated peaks is excluded from further analysis [23]. Data processing usually involves preprocessing steps (e.g., normalization and mean-centering) and some kind of scaling method (e.g., auto, root mean, filter, or shift variance scaling) [17]. Score plots and loading plots are commonly used for data visualization [24]. With such approach it has been possible with ToF-SIMS to differentiate films made of different proteins [2, 22], characterize extracellular matrix scaffolds [20, 25] or to separate human cell and bacterium "footprints" [6].

2 Materials

For all kinds of samples: phosphate buffered saline (PBS), deionized water (dH_2O), sample substrates (*see* **Note 1**). For frozen samples: a cryosectioning device, a freeze drying apparatus. For formalin-fixed and paraffin-embedded samples: microtome, oven, xylene, and ethanol. For extracellular matrix samples: trypsin, trypsin neutralization solution. If vacuum drying is preferred, a vacuum drying machine is needed. If the samples are kept under vacuum during storage, for example a vacuum desiccator can be used. For sample analysis, a ToF-SIMS machine is needed.

3 Methods

3.1 Sample Processing for Protein Samples

1. Rinse the protein coated samples three times with PBS.
2. Wash three times with dH_2O during 3 min to remove buffer salts (*see* **Note 2**).
3. Dry the samples. In this step air drying or vacuum drying should be performed. If samples are air-dried place them in a vertical position in order to avoid liquid drops formation over the sample (*see* **Note 3**).

3.2 Sample Processing for Cell Samples

1. Culture the cells over the sample substrate following the specific cell culture protocol. Be sure that the cells are able to adhere to the sample substrate (*see* **Note 4**).
2. When confluence is achieved take the sample substrate from the culture system.
3. Wash three times with PBS during 3 min.
4. Wash three times with PBS dilutions in dH_2O (72:25; 50:50; 25:75) and finally wash one time in dH_2O (*see* **Note 5**).
5. Dry the samples. In this step air drying or vacuum drying should be performed. If samples are air-dried place them in a vertical position in order to avoid liquid drops formation over the sample (*see* **Note 3**).
6. Take the samples for ToF-SIMS analysis (*see* **Note 6**).

3.3 Sample Processing for Frozen Samples

1. Cut the frozen section samples to 10 or 4 μm thick sections. Frozen sections should be cut at −20 °C in a cryosectioning device. It is important that the glass slides (or any other type of sample holders used) are pre-cooled (*see* **Note** 7).
2. Place the samples in freeze dying apparatus for about 12 h (overnight) at 10^3 mbar.
3. After freeze drying procedure is finished keep the samples in freeze conditions (for cryosection samples) and under vacuum until ToF-SIMs analysis is performed. This is of primary importance to prevent molecular delocalization of non-fixable molecules, i.e., movement of molecules from the original position.
4. Take the samples for ToF-SIMS analysis. Note that that frozen cryosections should be kept in cold conditions, and therefore, the ToF-SIMS apparatus to be used should have cold stage holder conditions.

3.4 Sample Processing for Formalin-Fixed and Paraffin-Embedded (FFPE) Tissues

1. Cut the FFPE tissue to 4 μm thick sections and place them over the sample substrate. FFPE tissue should be cut in microtome at room temperature (≈23 °C).
2. Place samples in oven at 37 °C during overnight in order to achieve full tissue adherence. This is of importance specially if you are dealing with hard tissues, e.g., bone or cartilage.
3. Deparaffinize and rehydrate the FFPE samples following standard deparaffinization protocols (*see* **Note 8**).
4. Dry the samples. In this step air drying or vacuum drying should be performed. If sample are air-dried place them in a vertical position in order to avoid liquid drops formation over the sample (*see* **Notes 9** and **10**).
5. After drying procedure is finished keep samples under vacuum until ToF-SIMs analysis is performed. This is of primary importance in order to prevent surface contamination.

3.5 Sample Processing for Extracellular Matrix Samples

1. Culture the adherent cells over the sample substrate following the cell culture specific protocol. Be sure that the cells are able to adhere to the sample substrate.
2. When confluence is reached trypsinize the cells as usually (*see* **Note 11**).
3. Neutralize with trypsin neutralization medium and remove the medium.
4. Rinse the protein coated samples three times with PBS.
5. Wash three times with dH_2O during 3 min to remove buffer salts.
6. Dry the samples. In this step air drying or vacuum drying should be performed. If samples are air-dried place them in a vertical position in order to avoid liquid drops formation over the sample.

3.6 Protocol for ToF-SIMS

1. For tuning and calibration, please refer to International Organization for Standardization (ISO) standard 23830 (*see* **Note 12**).

3.7 Multivariate Analysis of the ToF-SIMS Data

Figure 1 is an example of two ToF-SIMS spectra showing the positive ion peaks.

1. *Peak selection* is necessary to focus the analysis on the variable of interest, if the set of samples being investigated have several properties that vary (substrates, type of proteins, surface coverage, protein conformation, protein orientation, etc.), because it is extremely challenging to separate the effect of the different variables using a single dataset. Furthermore, ToF-SIMS spectra contain many mass channels that contain only noise, or no counts, due to the low background between peaks.

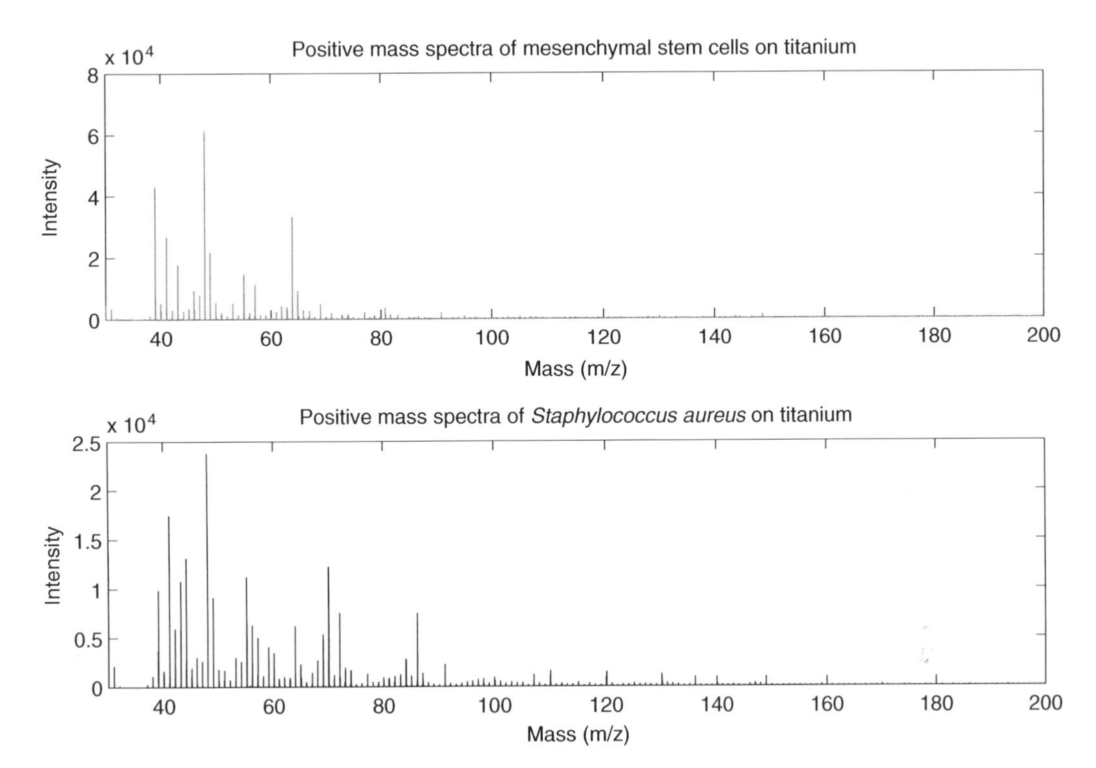

Fig. 1 An example of ToF-SIMS spectra showing the positive ion peaks

Therefore, multivariate analysis of proteins is commonly done using the peaks known to be related to amino acids (*see* **Note 13**). As an example, Fig. 2 shows closer view of tryptophan-related peaks of the same spectra that is show in Fig. 1.

2. *Data normalization* is done by dividing each variable (peak) in the matrix by a scalar value. Many issues affect how easily the molecules detach from the surface and this affects the intensity of the spectra (e.g., in Fig. 1, the scale of the *y*-axis is different between the two spectra). Data normalization is done to get different spectra on the same scale, which enables comparison of the differences between samples (*see* Fig. 3a). Common normalization methods include normalization to the total intensity of a given spectra, to the most intense peak in a given spectrum, and to the sum of all selected peaks from the spectrum (*see* **Note 14**).
3. *Mean-centering* is done by subtracting the average value of that variable in each spectrum. This results in a data set where each variable varies across a common mean of zero (*see* Fig. 3b). The aim of mean centering is to focus on the differences and not the similarities in the data.
4. *Scaling and transformation*. Autoscaling, root mean scaling, filter scaling, and shift variance scaling provide significant improvement of the unscaled data [17]. Scaling methods are

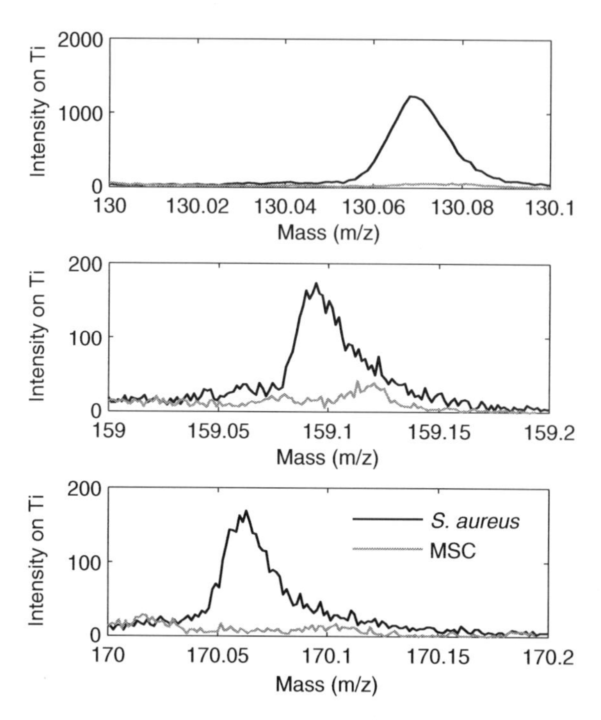

Fig. 2 Closer view of ToF-SIMS spectra of peaks that are typical for tryptophan (after data normalization)

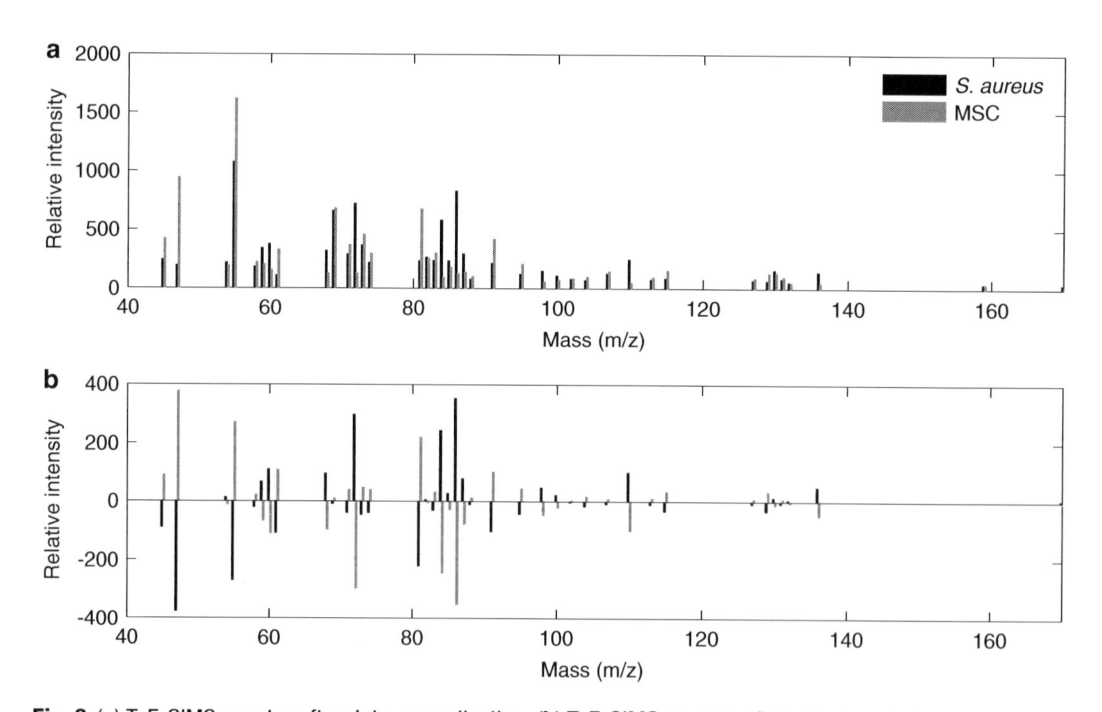

Fig. 3 (**a**) ToF-SIMS spectra after data normalization. (**b**) ToF-SIMS spectra after mean centering

data pretreatment approaches that divide each variable by a factor, the scaling factor, which is different for each variable. They aim to adjust for the differences in fold differences between the different analytes by converting the data into differences in concentration relative to the scaling factor. Data transformation (e.g., log transformation or power transformation) aims to correct for heteroscedasticity.

5. *Principal component analysis.* After appropriate preprocessing of the data, apply principal component analysis.

4 Notes

1. Suitable sample substrates are for example silicon substrates, metal coated silicon substrates, mica (a type of silicate) substrates, and coated glass slides. Providers of the ToF-SIMS machines are for example ION-TOF GmbH (Münster, Germany) and Physical Electronics (Chanhassen, MN, USA).
2. Some researchers place the samples in dH_2O for 24 h to ensure the removal of any remaining ions from the buffer [6, 26]. This is sometimes necessary to get rid of excess sodium, potassium, chlorine, etc.
3. Air drying will cause chemical rearrangements. It has been successfully applied for simple analysis, e.g., separation of different types of samples (mesenchymal stem cell from *Staphylococcus aureus* [6] or different breast cancer cell types from each other [5]). However, for more sophisticated analysis we recommend freeze drying. Furthermore, note that ethanol treatment may extract the lipids from the cells [27]. Therefore, for example critical point dehydration (where ethanol dehydration is a pre-requirement) is not recommended.
4. Follow cell culture guidelines and assure that the sample substrates are appropriately sterilized.
5. Washing with increase concentration of dH_2O until finally washing with just dH_2O prevents rupture of the cells given the fact that dH_2O is a hypo-osmotic solution. This is of considerable importance if cellular cytosol is to be analyzed.
6. When you prepare ToF-SIMS samples, the samples and the sample holders should be handled with clean tweezers and gloves. Please use polyethylene gloves rather than latex gloves as some latex gloves contain silicones. Silicone is one of the most common contaminants of surface.
7. Adherence between the cut frozen section and glass slide might be particularly difficult. One possibility to improve adherence can be by contacting the backside of the substrate

with fingers until it warm ups and the tissue frozen sections just start to thaw and taking the sample immediately after that back to frozen conditions. Cryosectioning machine can be purchased from different companies. Leica is a common brand among Research Institutes.

8. If lipids are to be included in ToF-SIMS analysis this procedure should be avoided, given that the use of alcohol effectively dissolves and extracts lipids out of the samples during such procedures. Frozen sections are preferred to such analysis.
9. If liquid drops are allowed to air-dry over your sample, this will be visible in the ToF-SIMS spectrum given the increase residuum concentration from the medium that forms around the cells.
10. Some researchers apply a controlled stream of argon gas directly to the sample placed in vertical position to allow the sample to dry faster [28].
11. Trypsinization does some damage to the cells and extracellular matrix. To preserve the extracellular matrix better, ppNIPAM, a "smart polymer," can be used as a less-destructive method of cell monolayer harvest [25].
12. Collecting data from at least 3–5 spots per sample across at least two samples for homogeneous surfaces, and 5–7 spots per sample across at least three samples for non-homogeneous samples for ToF-SIMS spectra is suggested [29].
13. Due to the large number of peaks typically seen at each nominal mass in ToF-SIMS spectra from organic and biological surfaces, manual peak selection is recommended [28]. To help the interpretation of the data, some general rules are presented [30]. First, at very low mass ranges, (for example, below m/z 25), a series of background peaks characteristic of carbon (C), oxygen (O), and hydrogen (H) are observed. Second, with a few exceptions, inorganic species have a mass defect, i.e., the exact mass of these peaks observed in the mass spectrum is less than the nominal mass. In contrast, organic species have a mass excess, i.e., the exact mass of the peaks is greater than the nominal mass. Third, SIMS organic fragments and molecular ions are dominated by even electron ions, whereas carbon-based species typically results in odd nominal masses. Fourth, electronegative species produce intense atomic ions in the negative ion mode and conversely with electropositive species. Peaks known to be related to amino acids are listed for example in refs. 15, 16, 20, 21.
14. Normalization is done to remove variance in the data that is due to differences such as sample charging, instrumental conditions, or topography.

References

1. Brulet M, Seyer A, Edelman A, Brunelle A, Fritsch J, Ollero M, Laprévote O (2010) Lipid mapping of colonic mucosa by cluster TOF-SIMS imaging and multivariate analysis in cftr knockout mice. J Lipid Res 51:3034–3045
2. Wagner MS, Shen M, Horbett TA, Castner DG (2003) Quantitative analysis of binary adsorbed protein films by time of flight secondary ion mass spectrometry. J Biomed Mater Res A 64:1–11
3. Wagner MS, Horbett TA, Castner DG (2003) Characterizing multicomponent adsorbed protein films using electron spectroscopy for chemical analysis, time-of-flight secondary ion mass spectrometry, and radiolabeling: capabilities and limitations. Biomaterials 24: 1897–1908
4. Jungnickel H, Jones EA, Lockyer NP, Oliver SG, Stephens GM, Vickerman JC (2005) Application of TOF-SIMS with chemometrics to discriminate between four different yeast strains from the species Candida glabrata and Saccharomyces cerevisiae. Anal Chem 77: 1740–1745
5. Kulp KS, Berman ES, Knize MG, Shattuck DL, Nelson EJ, Wu L, Montgomery JL, Felton JS, Wu KJ (2006) Chemical and biological differentiation of three human breast cancer cell types using time-of-flight secondary ion mass spectrometry. Anal Chem 78: 3651–3658
6. Kaivosoja E, Virtanen S, Rautemaa R, Lappalainen R, Konttinen YT (2012) Spectroscopy in the analysis of bacterial and eukaryotic cell footprints on implant surfaces. Eur Cell Mater 24:60–73
7. Wu L, Lu X, Kulp K, Knize M, Berman E, Nelson E, Felton J, Wu KJ (2007) Imaging and differentiation of mouse embryo tissues by ToF-SIMS. Int J Mass Spectrom 260: 137–145
8. Barnes CA, Brison J, Michel R, Brown BN, Castner DG, Badylak SF, Ratner BD (2011) The surface molecular functionality of decellularized extracellular matrices. Biomaterials 32:137–143
9. Cillero-Pastor B, Eijkel G, Kiss A, Blanco FJ, Heeren RM (2012) Time-of-flight secondary ion mass spectrometry-based molecular distribution distinguishing healthy and osteoarthritic human cartilage. Anal Chem 84:8909–8916
10. Frisz JF, Choi JS, Wilson RL, Harley BA, Kraft ML (2012) Identifying differentiation stage of individual primary hematopoietic cells from mouse bone marrow by multivariate analysis of TOF-secondary ion mass spectrometry data. Anal Chem 84:4307–4313
11. Werner H (2003) SIMS: from research to production control. Surf Interface Anal 35:859–879
12. Benninghoven A, Rudenauer FG, Werner HW (1987) Secondary ion mass spectrometry: basic concepts, instrumental aspects, applications and trends. Wiley-Interscience, New York
13. de Hoffmann E, Stroobant V (2001) Mass spectrometry: principles and applications. Wiley, New York
14. Vickerman JC, Brown A, Reed NM (1989) Secondary ion mass spectrometry: principles and applications. Clarendon Press
15. Chait B, Standing K (1981) A time-of-flight mass spectrometer for measurement of secondary ion mass spectra international. J Mass Spectrom Ion Phys 40:185–193
16. Samuel NT, Wagner MS, Dornfeld KD, Castner DG (2001) Analysis of poly-amino acids by static time-of-flight secondary ion mass spectrometry TOF-SIMS. Surf Sci Spectra 8:163–184
17. Wagner MS, Castner DG (2001) Characterization of adsorbed protein films by time-of-flight secondary ion mass spectrometry with principal component analysis. Langmuir 17:4649–4660
18. Tyler BJ, Rayal G, Castner DG (2007) Multivariate analysis strategies for processing ToF-SIMS images of biomaterials. Biomaterials 28:2412–2423
19. Jackson JE (1980) Principle components and factor analysis: part I—principal components. J Qual Tech 12:201–213
20. Jackson JE (1991) A user's guide to principal components. Wiley, New York
21. Canavan HE, Graham DJ, Cheng X, Ratner BD, Castner DG (2007) Comparison of native extracellular matrix with adsorbed protein films using secondary ion mass spectrometry. Langmuir 23:50–56
22. Henry M, Bertrand P (2009) Surface composition of insulin and albumin adsorbed on polymer substrates as revealed by multivariate analysis of ToF-SIMS data. Surf Interface Anal 41:105–113
23. Lhoest JB, Wagner MS, Tidwell CD, Castner DG (2001) Characterization of adsorbed protein films by time of flight secondary ion mass spectrometry. J Biomed Mater Res 57:432–440
24. Wagner MS, McArthur SL, Shen M, Horbett TA, Castner DG (2002) Limits of detection for time of flight secondary ion mass spectrometry (ToF-SIMS) and X-ray photoelectron spectroscopy (XPS): detection of low amounts of adsorbed protein. J Biomater Sci Polym Ed 13:407–428

25. Brown BN, Barnes CA, Kasick RT, Michel R, Gilbert TW, Beer-Stolz D, Castner DG, Ratner BD, Badylak SF (2010) Surface characterization of extracellular matrix scaffolds. Biomaterials 31:428–437
26. Canavan HE, Cheng X, Graham DJ, Ratner BD, Castner DGJ (2005) Cell sheet detachment affects the extracellular matrix: a surface science study comparing thermal liftoff, enzymatic, and mechanical methods. J Biomed Mater Res A 75:1–13
27. Denault L, Ostrowski SG, Smentkowski VS, Paxon TL (2008) Comparison of sample preparation methods for analysis of biological samples by high vacuum techniques. Microsc Microanal 14(S2):28–29
28. Berman ES, Fortson SL, Checchi KD, Wu L, Felton JS, Wu KJ, Kulp KS (2008) Preparation of single cells for imaging/profiling mass spectrometry. J Am Soc Mass Spectrom 19:1230–1236
29. Graham DJ, Castner DG (2012) Multivariate analysis of ToF-SIMS data from multicomponent systems: the why, when, and how. Biointerphases 7:49
30. E2695-09 Standard guide for interpretation of mass spectral data acquired with time-of-flight secondary ion mass spectroscopy. ASTM International

Chapter 16

Label-Free Imaging of Adipogenesis by Coherent Anti-Stokes Raman Scattering Microscopy

Antti Isomäki, Tarvo Sillat, Mari Ainola, Mikko Liljeström, Yrjö T. Konttinen, and Mika Hukkanen

Abstract

Label-free imaging technologies to monitor the events associated with early, intermediate and late adipogenic differentiation in multipotent mesenchymal stromal cells (MSCs) offer an attractive and convenient alternative to conventional fixative based lipid dyes such as Oil Red O and Sudan Red, fluorescent labels such as LipidTOX, and more indirect methods such as qRT-PCR analyses of specific adipocyte differentiation markers such as peroxisome PPARγ and LPL. Coherent anti-Stokes Raman scattering (CARS) microscopy of live cells is a sensitive and fast imaging method enabling evaluation of the adipogenic differentiation with chemical specificity. CARS microscopy is based on imaging structures of interest by displaying the characteristic intrinsic vibrational contrast of chemical bonds. The method is nontoxic, non-destructive, and minimally invasive, thus presenting a promising method for longitudinal analyses of live cells and tissues. CARS provides a coherently emitted signal that is much stronger than the spontaneous Raman scattering. The anti-Stokes signal is blue shifted from the incident wavelength, thus reducing the non-vibrational background present in most biological materials. In this chapter, we aim to provide a detailed approach on how to induce adipogenic differentiation in MSC cultures, and present our methods related to label-free CARS imaging of the events associated with the adipogenesis.

Key words Adipogenesis, Adipocyte, CARS microscopy, Label-free imaging, Lipids, Live cell analysis, Mesenchymal stromal cell

1 Introduction

Adipocytes and adipogenesis have assumed significant relevance due to the detection of adipokines and hormones secreted by the adipose tissue including cell-to-cell signalling leptin, adiponectin, resistin, and visfatin, among many others. These hormones and adipokines have recognized roles in several metabolic disorders and diseases such as obesity, metabolic syndrome, type II diabetes, cardiovascular disease, and various degenerative and inflammatory diseases affecting bone metabolism [1], cartilage homeostasis [2], and autoimmune diseases such as rheumatoid arthritis [3].

Shunichi Shiozawa (ed.), *Arthritis Research: Methods and Protocols*, Methods in Molecular Biology, vol. 1142, DOI 10.1007/978-1-4939-0404-4_16,

In order to better understand the adipogenesis, such cells are differentiated from the multipotent mesenchymal stromal cells (MSCs) using adipogenic medium that consists of dexamethasone, indomethacin, insulin, and isobutylmethylxanthine. The differentiation process is conventionally monitored using early, intermediate, and late differentiation markers, e.g., peroxisome proliferator-activated receptor gamma (PPARγ), lipoprotein lipase (LPL), and leptin in addition to the detection of the forming lipid droplets with Oil Red O staining [4]. However, the use of these conventional markers requires extensive processing of the samples, whereas imaging of adipogenesis by Coherent anti-Stokes Raman scattering (CARS) microscopy can be utilized in live cell cultures without prior sample processing or specific labelling of the forming intracellular lipid droplets.

CARS microscopy is an emerging technique with some unique properties. It provides label-free information with chemical specificity from biological samples distinguishing many biopolymers such as nucleic acids, lipids, proteins, and carbohydrates based on their specific molecular vibrations. Since the CARS signal arises as a result of a nonlinear multiphoton process using near-infrared lasers, it naturally provides the ability of optical sectioning with minimal phototoxicity, and with a resolution comparable to conventional confocal microscopy techniques. There is an increasing number of investigations where CARS has been used for studies of relevance to adipogenesis [5, 6]. In particular, the strong Raman signals arising from CH_2 groups of saturated lipids are commonly exploited making CARS microscopy particularly suited to the studies of adipogenesis [7]. These investigations can be further extended to the studies of lipid domains, trafficking, storage and metabolism [8–11], in addition to pharmacological approaches where CARS imaging has been utilized as a complementary tool to study the effects of statins on protein lipidation [12]. The potential of combining CARS microscopy with other nonlinear imaging modalities, such as second harmonic generation (SHG) and multiphoton (MP) microscopy is also a promising development in deep tissue imaging. These imaging modalities have been combined in a multimodal label-free nonlinear optical imaging study of the development of atherosclerotic plaques in myocardial infarction-prone animal model in which CARS imaging was used to highlight the lipid rich fat plagues, SHG to illuminate the fibrillar collagen, and MP to demonstrate the elastin fibers within aortic tissues [13].

In this chapter, we provide a detailed approach and outline on the procedures related to the sequential differentiation of MSCs to mature adipocytes, and the key aspects concerning label-free CARS imaging of the adipogenesis in live cells undergoing the adipogenic differentiation.

2 Materials

2.1 Isolation of Bone Marrow Multipotent Mesenchymal Stromal Cells

Ficoll-Paque, GE Healthcare Biosciences, Uppsala, Sweden.

2.2 Cell Culture and Induction of Adipogenesis

1. Basal medium: DMEM-low glucose (Gibco, Life Technologies, Paisley, UK), fetal bovine serum (StemCell Technologies, Vancouver, Canada), penicillin/streptomycin (Gibco, Life Technologies).
2. Dexamethasone, isobutylmethylxanthine, indomethacin (all from Sigma-Aldrich, St. Louis, MO), insulin (Actrapid Penfil 100 IU/ml, Novo Nordisk, Espoo, Finland).

2.3 Oil Red O and DAPI Staining

1. Distilled H_2O, phosphate buffered saline (PBS; Invitrogen), paraformaldehyde (BDH).
2. Oil Red O (Sigma-Aldrich O-0625), isopropanol (2-propanol; Sigma-Aldrich), water bath, distilled H_2O, and syringe filters with 0.45 μm pore size (33 mm Millex Filter Units, Merck Millipore, Billerica, MA).
3. 4′,6-diamidino-2-phenylindole (DAPI, Vector Laboratories, Burlingame, CA), Fluoro-Gel mounting medium (EMS, Hatfield, PA).
4. Olympus AX70 Provis epifluorescence microscope (Olympus, Hamburg, Germany) with a suitable filter block to detect DNA specific DAPI signal (e.g., excitation 350/50, beam splitter 400 LP, emission 460/50; Chroma 49000-ET-DAPI).

2.4 Quantitative Real Time-Polymerase Chain Reaction (qRT-PCR)

1. RNeasy Mini Kit (Qiagen, Hilden, Germany).
2. Spectrophotometry: NanoDrop ND-1000 (Thermo Scientific, Wilmington, DE, USA).
3. iScript cDNA synthesis kit (Bio-Rad Laboratories, Hercules, CA).
4. iQ SYBR Green supermix (Bio-Rad Laboratories).
5. Primer sequences, GenBank accession numbers, and corresponding amplicon lengths: peroxisome proliferator-activated receptor gamma (PPARγ) (GenBank: NM_138712) forward 5′-GTGAAGGATGCAAGGGTTTC-3′ and reverse 5′-TCAGCGGGAAGGACTTTATG-3′, length 303 bp; lipoprotein lipase (LPL) (GenBank: NM_000237) forward 5′-CCGGTTTATCAACTGGATGG-3′ and reverse 5′-AATCACGCGGATAGCTTCTC-3′, length 349 bp; leptin (LEP) (Genbank: NM_000230) forward 5′-AACCCTGTGCGGATTCTTGT-3′ and reverse 5′-CCAGGTCGTTGGATATTTGG-3′, length 293 bp; ribosomal protein, large, P0 (RPLP0) (GenBank: NM_001002) forward 5′-GGCGACCTGGAAGTCCAACT-3′ and reverse

5′-CCATCAGCACCACAGCCTTC-3′, length 149 bp (*see* **Note 1**). For primers, corresponding sequences are searched from NCBI Entrez search system and designed and verified with NCBI primer designing tool (Primer-BLAST) after which the primers are produced by Oligomer Oy (Helsinki, Finland) [14].

6. iQ5 PCR machine (Bio-Rad Laboratories).

2.5 Confocal CARS Microscope

Our commercial confocal CARS microscope is based on a Leica DMI 6000 CS inverted microscope and Leica TCS SP8 CARS system that combines confocal, multiphoton (MP), second harmonic generation (SHG), and CARS modalities.

General components

1. Lasers for conventional confocal imaging: diode 405 nm; argon laser: 458, 476, 488, 496, 514 nm; DPSS-laser: 561 nm; HeNe-laser: 594, 633 nm.
2. Tandem scanner with both point and high speed resonant scanning (switchable between a large field-of-view point scanner and a fast resonant scanning system for live cell imaging).
3. Two hybrid GaAsP detectors and one standard PMT for confocal detection.
4. Motorized stage for multipoint scanning, galvanometric Z-stage, hardware autofocus.
5. Ludin full-enclosure black environmental chamber for temperature, CO_2 (5 %), and O_2 (0.5–25 %) controls. Water micro dispenser for automated supply of water immersion during the experiment.
6. HCX IRAPO L 25×/0.95 (water immersion), HC PL APO 63×/1.20 (water immersion), and HC PL APO 63×/1.30 (glycerol immersion) objectives.
7. Anti-vibration optical table.
8. Leica LAS AF (version 3.1.0) software for microscope controls and image acquisition.

Multiphoton components:

9. The system is equipped with a PicoEmerald solid-state near infrared laser (APE Angewandte Physik & Elektronik GmbH, Berlin, Germany) that produces both the pump and the Stokes beams necessary for the CARS process. The average power of both beams is ~1.3 W. The pulse frequency is 80 MHz and the pulse duration is 5–6 ps corresponding to the bandwidth of 2–3/cm^{-1}. The Stokes wavelength is fixed to 1,064.5 nm and the pump wavelength can be tuned from 780 to 940 nm. Thus, the accessible probing range is from 1,250 to 3,400/cm^{-1}. This range covers several biologically relevant vibrational frequencies some of which are listed in Table 1.

Table 1
Examples of Raman-active vibrational bands of lipids and related species

Vibrational mode	Raman shift (cm^{-1})	Species	References
C–H_2 stretch	2,845	Saturated fatty acids	[10]
C=C–H, C=C stretch	1,270, 1,660, 3,015, 3,050	Unsaturated fatty acids	[10]
C–H deformation C=C stretch	1,440 1,674	Cholesterol	[15]
C=O	1,739	Ester group in phospholipids	[15]

10. Four external non-descanned detectors (NDDs) are used for collecting the signals in the forward (F) and backward (epi) scattered directions. The following detection ranges are set by the emission filters:
    - F-CARS and Epi-CARS: 560–750 nm.
    - F-SHG and Epi-SHG: 380–550 nm.

    The relatively broad detection ranges are needed to allow for the tunability of the CARS signal.

2.6 Data Analysis

1. Imaris 7.6.1 (Bitplane AG, Zurich, Switzerland) software for visualizing, processing and analyzing of the images.

3 Methods

3.1 Isolation of Bone Marrow Multipotent Mesenchymal Stromal Cells

1. Mesenchymal stromal cells are obtained from bone marrow aspirates from healthy young donors (*see* **Notes 2** and **3**).
2. 10–20 ml of bone marrow is aspirated under local anesthesia from the posterior iliac crest and collected into heparinized tubes.
3. Mononuclear cells are isolated with a density gradient (Ficoll-Paque, GE Healthcare Biosciences, Uppsala, Sweden) and plated at $4 \times 10^5/cm^2$ on a cell culture dish in basal medium (low-glucose DMEM (Invitrogen, Paisley, UK)) with 10 % MSC-Qualified FBS (fetal bovine serum, StemCell Technologies, Vancouver, Canada) and antibiotics.
4. After 72 h the non-adherent cells are discarded and the plates are thoroughly washed with phosphate-buffered saline (PBS). The basal medium on adherent cells is changed twice a week until a confluence of 80 % is achieved whereupon the cells are harvested by incubation in PBS containing trypsin and EDTA for 5–10 min and thereafter replated at 1,000 cells/cm^2 (passage 1).

3.2 Verification of the Mesenchymal Phenotype

Flow cytometry is also often used for the verification of the cell phenotype. Mesenchymal stromal cells should be positive for CD29, CD44, CD105, CD166, and negative for CD14, CD34, and CD45. For these analyses we use FACSCalibur (Beckton Dickinson, San Jose, CA) and Cell Quest Pro software (Beckton Dickinson) (*see* **Note 4**).

3.3 Cell Culture and Induction of Adipogenesis

1. The cells are cultured in 10 cm petri dishes in low-glucose DMEM (Invitrogen, Paisley, UK) with 10 % MSC-Qualified FBS (StemCell Technologies, Vancouver, Canada).
2. The appropriate initial seeding density is 5×10^3 cells per cm^2.
3. Add the calculated volume of cell suspension to each dish and gently rock to disperse the cell suspension over the growth surface. Avoid excessive up and down pipetting.
4. Allow the mesenchymal cells to adhere and spread for 24–36 h in standard cell culture incubator with 37 °C, 5 % CO_2, and 90 % humidity.
5. Wash thoroughly with phosphate-buffered saline (PBS, pH 7.4) to remove all non-adherent cells.
6. Change the basal medium every 3–4 days.
7. After 3 weeks, when the cultures are in near confluence, the cells are passaged at a density of 5×10^3 cells per cm^2.
8. The cells are again allowed to proliferate and re-passaged as above in near confluence.
9. For adipogenic differentiation, cells from passage 4–5 are plated into cover-glass bottom multiwell chambers (Lab-Tek II, Thermo Scientific) or cover-glass bottom petri dishes (MatTek) (for inverted microscope designs). The appropriate cell density is 20×10^3 cells per cm^2. 6-well plates are convenient for experiments involving cell extractions such as qRT-PCR.
10. Before starting the adipogenic differentiation, the cells are grown to full confluence.
11. The adipogenic induction medium consists of the basal medium, supplemented with 10 μg/ml insulin, 1 μM dexamethasone, 0.5 mM isobutylmethylxanthine, and 60 μM indomethacin. Adipogenic maintenance medium consists of basal medium supplemented with 10 μg/ml insulin.
12. The cells are first kept in induction medium for 72 h and then changed to maintenance medium for 24 h. Such induction and maintenance cycle is repeated up to 28 days of differentiation.

13. Intracellular lipid droplets are usually first seen to appear at day 7 of adipogenic differentiation, followed by a progressive increase in their size and number so that by day 28 the lipid droplets fill most of the cytoplasmic volume of the adipocytes.

3.4 Oil Red O Staining

1. This method is convenient for quick general screening of the progression of adipogenesis. Cells are plated at a density of 20×10^3 cells per cm^2 into Petri dishes with coverslips at the bottom.
2. Before starting the adipogenic differentiation, the cells are grown to full confluence.
3. Cells from different stages of adipogenesis (days 1, 7, 14, 21, and 28) are rinsed with PBS and fixed with 4 % paraformaldehyde at room temperature for 20 min.
4. Cells are washed with distilled PBS, and stored at 4 °C prior to Oil Red O staining.
5. Stock solution is prepared by dissolving 1 mg/ml Oil Red O (Sigma O-0625) in isopropanol and heated in water bath 1 h at 56 °C. Store the stock at room temperature. Working solution is prepared by mixing constantly 5 min 6 parts of Oil Red O stock and 4 parts of dH_2O.
6. Leave to sit at room temperature for 5 min prior to use, filter through a 0.45 μm filter and use within 1 h.
7. After 10 min of staining the cells are immediately washed 4 times with dH_2O.
8. Cell nuclei are counterstained for 5 min in 1 μg/ml 4′,6-diamidino-2-phenylindole (DAPI, Vector Laboratories, Burlingame, CA).
9. The preparations are rinsed with dH_2O and visualized under microscopy without embedding or they are mounted with aqueous mounting media (Fluoro-Gel, EMS, Hatfield, PA). Lipid droplets are stained red with Oil Red O when examined with transmitted light microscopy (Fig. 1a), cell nuclei are seen in light blue when examined with epifluorescence microscopy (excitation maxima 358 nm, emission maxima 461 nm).

3.5 Quantitative Real Time-Polymerase Chain Reaction (qRT-PCR)

1. Most of our investigations involve identification and quantification of target molecules from the RNA and protein extracts. Thus it is also convenient to analyze the progression of adipogenesis by qRT-PCR from these extracts. PPARγ, LPL, and leptin are enriched in adipose tissues and they are differentially expressed during the distinct stages of adipogenic differentiation of mesenchymal stem cells.
2. Total RNA is isolated from the cells undergoing adipogenic differentiation with RNeasy Mini Kit.

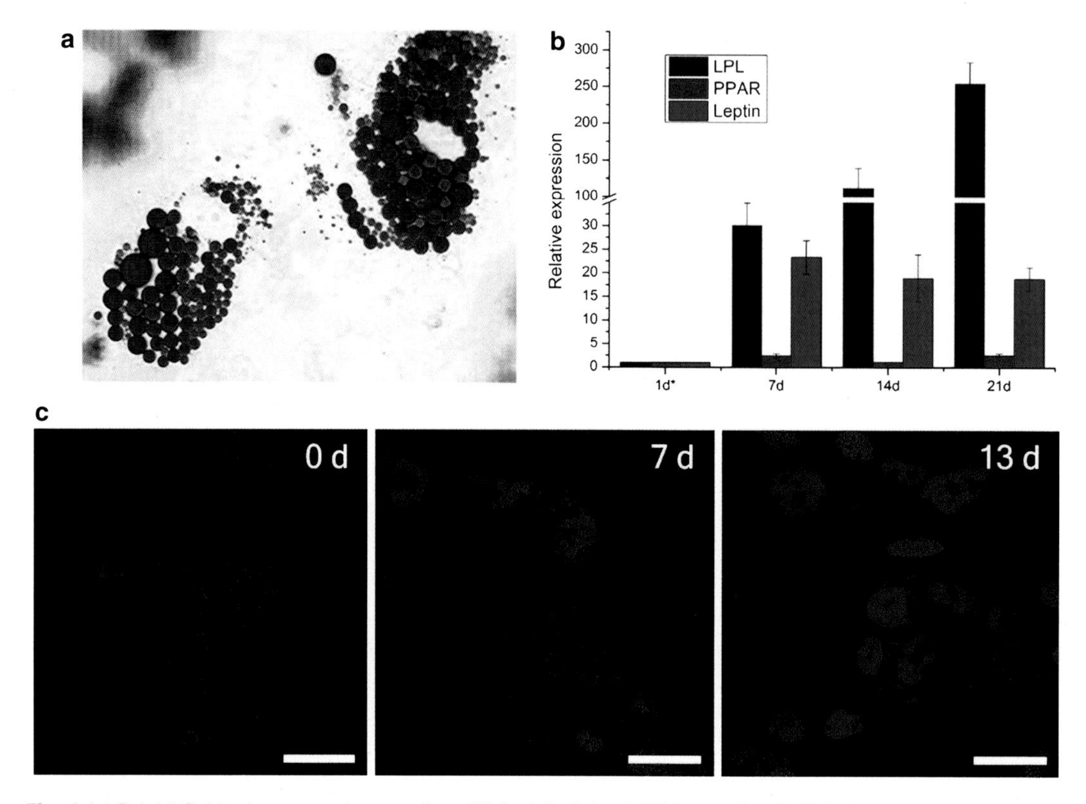

Fig. 1 (**a**) Bright-field microscope image of an Oil Red O stained MSC sample. (**b**) Relative gene expression of adipogenic markers in differentiated MSCs. mRNA was isolated at the indicated days of differentiation and relative gene expression of the adipogenic differentiation markers were analyzed by quantitative reverse transcription polymerase chain reaction. Results are normalized to expression of the housekeeping gene RPLPO and expressed relative to day 1 of differentiation. The results are means of five measurements and error bars represent SEM. (**c**) CARS time-series of unstained MSCs showing the progression of adipogenic differentiation during the first 13 days of the induction period. Scale bar is 100 μm

3. Total RNA concentration and quality is measured using the NanoDrop ND-1000 spectrophotometer (*see* **Note 5**) and complementary DNA (cDNA) is synthesized in 20 μl from 100 ng tRNA using iScript cDNA synthesis kit with a mixture of oligo (dT) and random hexamers to prime first-strand and RNase H+ iScript reverse transcriptase for cDNA synthesis (*see* **Note 6**).
4. Quantitative PCR is run on 2 μl of first strand cDNA using 250 nM primers in iQ SYBR Green supermix by iQ5 PCR machine. Each individual sample is amplified at least three times for all genes of interest.
5. qRT-PCR shows progressive increase in PPARγ and LPL gene expression during the adipogenic differentiation. PPARγ, LPL, and leptin mRNA levels are increased already at day 14 confirming the adipogenic differentiation (Fig. 1b) [4].

3.6 Confocal CARS Microscopy

1. CARS microscopy can be used for both fixed and live cells. This method describes the steps needed to follow the growth of the cells over time.
2. The prepared cell cultures are taken out of the cell culture incubator only for a short period of time (max 1 h) needed for imaging. The temperature in the environmental chamber of the microscope is set to 37 °C.
3. The cells are imaged within the maintenance medium in the cover-glass bottom multiwell chambers (Lab-Tek II, Thermo Scientific). The cover of the multiwell chamber can be kept in place during imaging.
4. Use an infrared corrected objective (here Leica HCX IRAPO L 25×/0.95) with water immersion to ensure optimal focusing of the CARS excitation light. The 63×/1.20 water and 63×/1.30 glycerol immersion objectives can be used for higher magnification, but with suboptimal focus correction and lower transmission in the IR spectral range.
5. Focus to the sample and locate the region of interest. Write down the coordinates of the region and a fixed reference point. A motorized XY-stage can be used to locate the same region in each of the time points (*see* **Note 7**).
6. Set the pump laser wavelength to 817 nm (Raman shift at 2,845/cm^{-1}). The Stokes laser wavelength is fixed at 1,064.5 nm (*see* **Note 8**).
7. Turn on and set the gain of the epi-directed and forward-directed PMTs to observe the signals in different detection geometries (*see* **Note 9**).
8. Set the Z-scan range to cover the full volume of the LDs within the region of interest.
9. Acquire the Z-stacks using the F-CARS channel for lipid imaging and the epi-SHG channel for the fluorescent signal (*see* **Note 10**). As an example, Fig. 1c shows the progression of adipogenic differentiation during the first 13 days of the induction period. Figure 2 shows the high resolution autofluorescence and CARS images of the mature adipocytes (day 21).
10. Return the sample to the cell culture incubator.

3.7 Data Analysis

The parameters of interest include the surface area, volume, and number of the lipid droplets in cells. An example of the analyzed volume distribution is shown in Fig. 3. When using additional fluorescent labels it is possible to relate this information to general cell morphology and localization of the organelles, e.g., the distance of the LDs from the nucleus.

Mechanisms leading to LD volume change can be analyzed using spatial distribution patterns measured at different time points of adipogenesis. As an example, analysis software can be used for tracking the movement and the fusion of LDs.

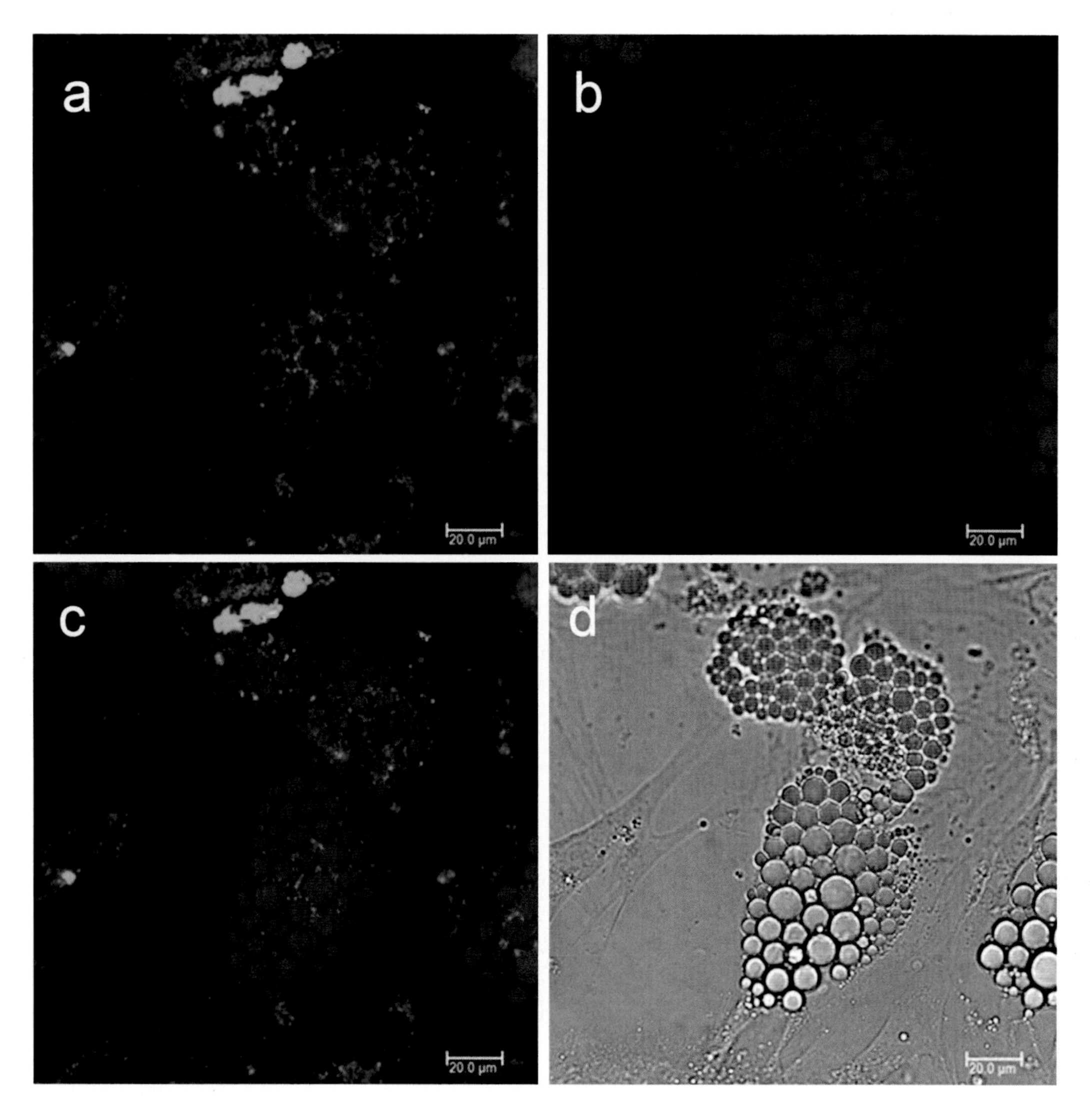

Fig. 2 Mature adipocytes on day 21 of adipogenesis. (**a**) Epi-SHG channel showing the autofluorescence of FAD rich areas and (**b**) F-CARS channel showing the lipid droplets at the CH_2 vibrational band (2,845/cm^{-1}). Images are maximum projections of a Z-stack. (**c**) Composite image of panels **a** and **b**. (**d**) Bright-field image of the same area. Scale bar is 20 μm

4 Notes

1. RPLPO is used as a housekeeping gene.
2. Commercially available human bone marrow-derived mesenchymal stem cells can be obtained for example from Lonza Group Ltd. (Human Poietics®, Lonza Group Ltd., Basel, Switzerland).

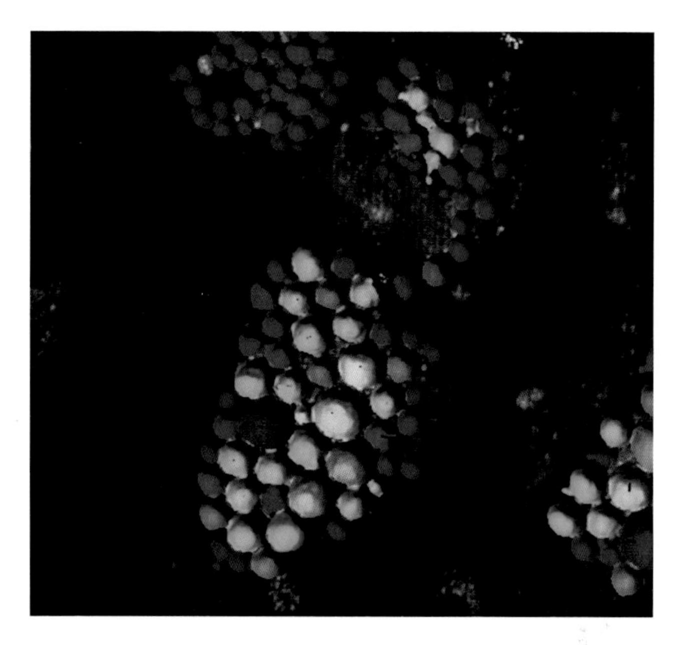

Fig. 3 Visualization of the volume distribution of the lipid droplets. The smallest LDs are shown in *violet* (the threshold for visualization was set to ~170 μm^3) and the largest (>2,300 μm^3) in *red color*

3. The capacity of mesenchymal stromal cells to proliferate and differentiate is dependent on the age of the donor and the number of cell passages. For these reasons, we prefer to use young age- and sex-matched donors, when feasible. Also, we usually prefer to use passages 4–5 for experimental purposes.
4. The International Society for Cellular Therapy has proposed the minimum standards for defining multipotent mesenchymal stromal cells. According to this statement, mesenchymal stromal cells must be plastic adherent under standard cell culture conditions. The cells should express surface markers CD73, CD90, and CD105, and lack the expression of CD11b or CD14, CD19, CD34, CD45, CD79a, and HLA-DR surface molecules. The cells must also be able to differentiate into osteoblasts, adipocytes and chondroblasts in vitro [16].
5. Total RNA is quantitated spectrophotometrically at OD260. Pure RNA has an OD_{260}/OD_{280} ratio around 2.0. A ratio of 1.8–2.0 corresponds to 90–100 % pure nucleic acid, which is fine.
6. If more than 100 ng tRNA is used, then appropriate dilution is made, e.g., 200 ng requires dilution 1:2, while 300 ng requires dilution 1:3.
7. The sample can be conveniently located by acquiring a bright-field image. Use a visible laser, e.g., 633 nm HeNe laser, and a transmitted light PMT in Scan-BF mode for detection.

When following specific cells over time a gridded cover-glass bottom dish helps in locating the same coordinates. This dish type is particularly useful if a motorized XY-stage is not available.

8. The CARS laser wavelengths are set to resonantly excite the vibrational state at the frequency of 2,845/cm^{-1}, corresponding to the strong Raman signal arising from CH_2 groups of saturated lipids. Scanning through a range of wave numbers may be necessary to find the Raman shift with the highest intensity.
9. Typically, for relatively large lipid droplets, the highest CARS signal is generated in the forward scattered direction and thus collected with the transmitted light PMT (F-CARS NDD channel). The two-photon excited fluorescence signal can be detected in the backward direction using the reflected light PMT (epi-SHG NDD channel) or the confocal detectors.
10. Fluorescent labels can be used in combination with label-free CARS imaging. Sequential measurement using standard visible light lasers and confocal detection together with CARS imaging is possible. Alternatively, the ps-pulses from the CARS laser can be used for two-photon excitation of the fluorescent markers (e.g., DAPI, CellTracker Green). The possible overlap between the fluorescent emission and the CARS signal should be taken into account in the selection of the fluorescent labels. Furthermore, the two-photon excitation can be used to image the intrinsic fluorescent species such as NAD(P)H and FAD in cells [17]. The ratio of NADH over FAD fluorescence can be used as an indicator of the cell metabolic state [18].

References

1. Cao JJ (2011) Effects of obesity on bone metabolism. J Orthop Surg Res 6:30
2. Dozio E, Corsi MM, Ruscica M, Passafaro L, Steffani L, Banfi G, Magni P (2011) Adipokine actions on cartilage homeostasis. Adv Clin Chem 55:61–79
3. Gómez R, Conde J, Scotece M, Gómez-Reino JJ, Lago F, Gualillo O (2011) What's new in our understanding of the role of adipokines in rheumatic diseases? Nat Rev Rheumatol 7:528–536
4. Sillat T, Saat R, Pöllänen R, Hukkanen M, Takagi M, Konttinen YT (2012) Basement membrane collagen type IV expression by human mesenchymal stem cells during adipogenic differentiation. J Cell Mol Med 16: 1485–1495
5. Evans CL, Xie XS (2008) Coherent anti-stokes Raman scattering microscopy: chemical imaging for biology and medicine. Annu Rev Anal Chem 1:883–909
6. Wang HW, Fu Y, Huff TB, Le TT, Wang H, Cheng JX (2009) Chasing lipids in health and diseases by coherent anti-Stokes Raman scattering microscopy. Vib Spectrosc 50:160–167
7. Nan X, Cheng JX, Xie XS (2003) Vibrational imaging of lipid droplets in live fibroblast cells with coherent anti-Stokes Raman scattering microscopy. J Lipid Res 44:2202–2208
8. Li L, Wang H, Cheng JX (2005) Quantitative coherent anti-Stokes scattering imaging of lipid distribution in coexisting domains. Biophys J 89:3480–3490
9. Le TT, Yue S, Cheng JX (2010) Shedding new light on lipid biology with CARS microscopy. J Lipid Res 51:3091–3102
10. Freudiger CW, Min W, Saar BG, Lu S, Holtom GR, He C, Tsai JC, Kang JX, Xie XS (2008) Label-free biomedical imaging with high sensitivity by stimulated Raman scattering microscopy. Science 322:1857–1861

11. Lyn RK, Kennedy DC, Sagan SM, Blais DR, Rouleau Y, Pegoraro AF, Xie XS, Stolow A, Pezacki JP (2009) Direct imaging of the disruption of hepatitis C virus replication complexes by inhibitors of lipid metabolism. Virology 394:130–142
12. Mörck C, Olsen L, Kurth C, Persson A, Storm NJ, Svensson E, Jansson JO, Hellqvist M, Enejder A, Faergeman NJ, Pilon M (2009) Statins inhibit protein lipidation and induce the unfolded protein response in the non-sterol producing nematode Caenorhabditis elegans. Proc Natl Acad Sci U S A 106:18285–18290
13. Ko AC, Ridsdale A, Smith MS, Mostaço-Guidolin LB, Hewko MD, Pegoraro AF, Kohlenberg EK, Schattka B, Shiomi M, Stolow A, Sowa MG (2010) Multimodal nonlinear optical imaging of atherosclerotic plaque development in myocardial infarction-prone rabbits. J Biomed Opt 15:020501
14. Ye J, Coulouris G, Zaretskaya I, Cutcutache I, Rozen S, Madden T (2012) Primer-BLAST: a tool to design target-specific primers for polymerase chain reaction. BMC Bioinformatics 13:134
15. Krafft C, Neudert L, Simat T, Salzer R (2005) Near infrared Raman spectra of human brain lipids. Spectrochim Acta A Mol Biomol Spectrosc 61:1529–1535
16. Dominici M, Le Blanc K, Mueller I, Slaper-Cortenbach I, Marini F, Krause D, Deans R, Keating A, Prockop D, Horwitz E (2006) Minimal criteria for defining multipotent mesenchymal stromal cells. The International Society for Cellular Therapy position statement. Cytotherapy 8:315–317
17. Zoumi A, Yeh A, Tromberg BJ (2002) Imaging cells and extracellular matrix in vivo by using second-harmonic generation and two-photon excited fluorescence. Appl Biol Sci 99:11014–11019
18. Zhuo S, Chen J, Luo T, Zou D (2006) Multimode nonlinear optical imaging of the dermis in ex vivo human skin based on the combination of multichannel mode and Lambda mode. Opt Express 14:7810–7820

Index

Shunichi Shiozawa (ed.), *Arthritis Research: Methods and Protocols*, Methods in Molecular Biology, vol. 1142, DOI 10.1007/978-1-4939-0404-4, © Springer Science+Business Media New York 2014

Q

R

S

T

V